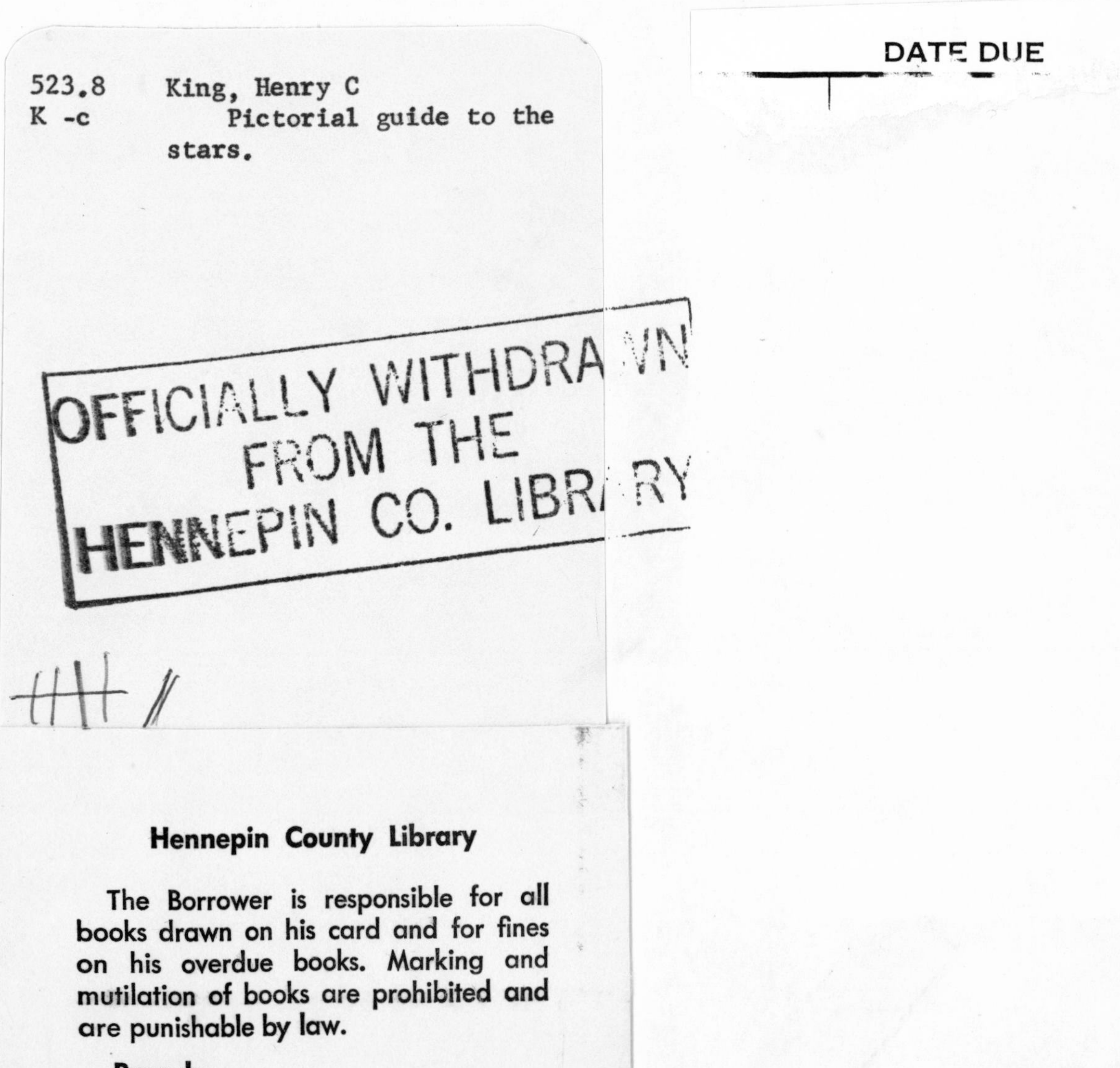

PICTORIAL GUIDE TO THE STARS

HENRY C.
KING

PICTORIAL GUIDE TO THE STARS

M20, the Trifid Nebula in Sagittarius. Photographed in red light.
COURTESY: MOUNT WILSON AND PALOMAR OBSERVATORIES

THOMAS Y. CROWELL COMPANY

ESTABLISHED 1834 / NEW YORK

To Frances Mary

Manufactured in the United States of America

L.C. Card 67–12404

1 2 3 4 5 6 7 8 9 10

FOREWORD

The object of this book is to give a reasonably clear and well-balanced account of modern knowledge about the stars. Most of the text is concerned with astrophysics, or the study of the nature of stars and interstellar material. This study is one of the most interesting parts of astronomy. Yet to the layman it can appear to be highly complex and abstruse, for its roots go deep into mathematics, chemistry, and physics. My aim has been to remove some of the complexity and mystery by making the treatment wholly descriptive and reducing technicalities to a minimum.

Wherever possible I have described the results of inquiries in astrophysics rather than the special tools and techniques used to obtain the results. To deal adequately with both of these aspects would have required not only a much larger book but also a more technical approach. The intellectual challenge and fascination of the subject lies almost solely in the data of observation and its interpretation. Most of the illustrations are concerned with data in the form of photographs of astronomical objects, while the text attempts to give the interpretation. If the study of the two together gives some idea of the nature of the astronomical universe and encourages the reader to pursue the subject further, my efforts will not have been in vain.

A major problem in writing and publishing books on astronomy is the great speed with which they tend to get out of date. A growing stream of new information obtained with optical and radio telescopes is constantly adding itself to that already assimilated by astronomers. Under its pressure, gaps in our knowledge are filled in, and time-honored ideas and theories may be swept away. The stream constantly disturbs the cosmological picture, or view of the universe and its larger parts, and our knowledge of the origin and evolution of the stars. But this is how it should be. No writer on astronomy, and certainly no astronomer, would expect or wish it to be otherwise.

Henry C. King

Toronto, Canada, 1967

CONTENTS

I	Introducing the Universe	1
II	Mapping the Stars	13
III	The Sun—Portrait of a Star	23
IV	The Sun's Nearest Neighbors	41
V	From Dwarfs to Supergiants	51
VI	Binary and Multiple Stars	62
VII	Exploding Stars	70
VIII	Pulsating Stars	77
IX	Stars and Nebulae	86
X	Star Clusters	102
XI	The Galaxy or Milky Way System	111
XII	Neighboring Galaxies	121
XIII	The Universe of Galaxies	132
XIV	The Expanding Universe	143
	Glossary	155
	Bibliography	160
	Index	161

CHAPTER I

INTRODUCING THE UNIVERSE

One of the most remarkable features of the starry sky is its rotation once in just under twenty-four hours. The stars move, or appear to move, as one great body in mainly an east–west direction. As they do so, they preserve their relative positions one with another, and so precisely as far as unaided vision is concerned that the night sky looks much the same over many centuries. It is as if the stars had no individual motions of their own, but instead were fixed to a rotating background, the sky itself.

The North Celestial Pole

The particular way in which the stars move depends on the geographical latitude of the observer. At the North Pole, for example, they travel around a point, known as the north celestial pole, which coincides with the zenith, or point directly overhead. It follows that they are all circumpolar, that is, they stay permanently above the horizon and disappear only in daytime. At the equator, on the other hand, the north celestial pole lies on the horizon at the north point, so none of the stars is circumpolar.

In mid-northern latitudes the north celestial pole lies directly on the celestial meridian, the imaginary line joining the zenith with the north point. Its precise altitude or elevation depends on the latitude of the place of observation. If the latter is 35 degrees, say, the north celestial pole has an altitude of 35 degrees. An observer at this latitude would therefore see as circumpolar those stars within a distance of 35 degrees of the north celestial pole. He would be able to see them on any clear night, but to see the rest would have to spread his observations over several months.

A good marker of the position of the north celestial pole is Polaris, the North Pole Star. It is fairly bright and with six other stars forms a pattern we know as the Little Dipper. Actually it is about one degree away from the celestial pole, but as far as unaided vision is concerned this difference is small enough to be ignored. Polaris, therefore, not only indicates the direction of true north but also, by the measure of its altitude, the latitude.

A similar situation is found in southern skies; the (south) celestial pole lies on that part of the celestial meridian between the zenith and the south point. No bright star indicates its position, but it lies a short distance from a faint naked-eye star known as Sigma Octantis. An observer in a mid-southern latitude therefore sees the circumpolar stars in the southern part of his sky, while one at the south pole itself sees all the stars as circumpolar.

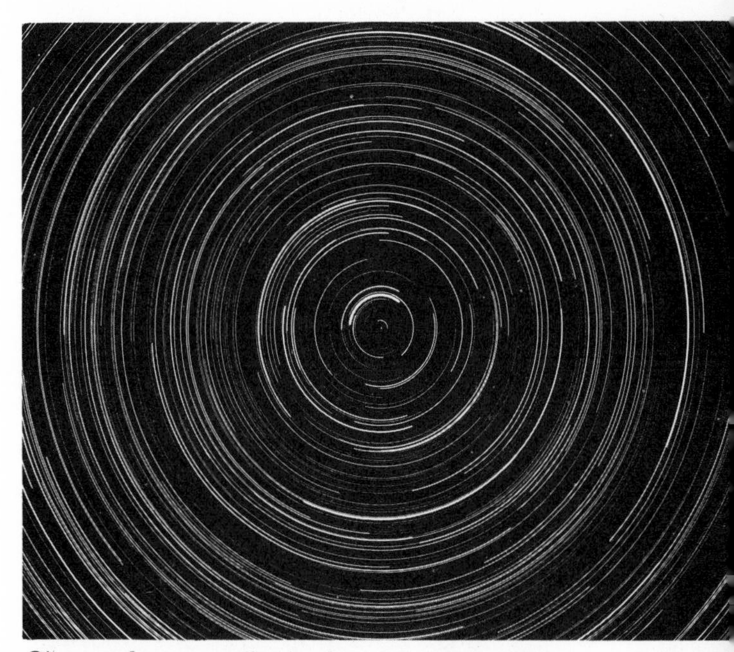

Circumpolar star trails. As the earth rotates, the northern stars appear to describe circles around the north celestial pole. This photograph was obtained by using a fixed camera and an exposure time of several hours.

—Lick Observatory

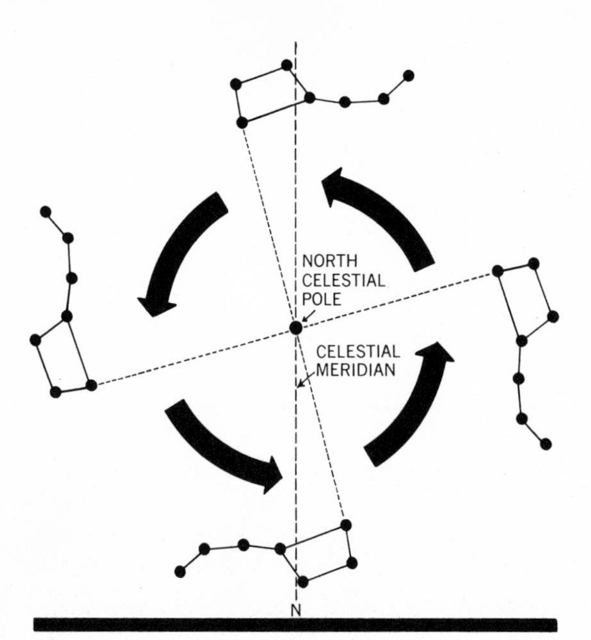

The north celestial pole and circumpolar stars. The Big Dipper appears to move around the north celestial pole once in about 23 hours 56 minutes 4 seconds. In the diagram the observer's latitude is taken as 40 degrees North. The angular height of the north celestial pole above the north point of the horizon is therefore 40 degrees, so the Big Dipper is circumpolar.

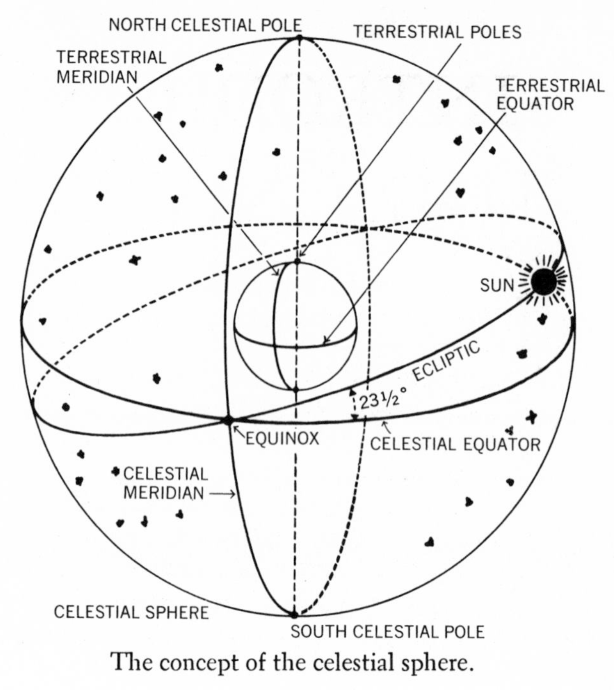

The concept of the celestial sphere.

Theories of the Universe

These variations of the night sky for different places led the Ancient Greeks to believe that the stars were fixed to an enormous sphere centered on the earth. The sphere rotated once a day on an axis which coincided with that of the earth. Its meridians, equator, and two poles were projections of the corresponding meridians, equator, and poles of the earth. It was, moreover, a *celestial* sphere, for its motion and form, like the nature of the stars attached to it, were supposed to be perfect, unchanging, eternal, and divine. The fixed central earth, on the other hand, was essentially noncelestial in the sense that it was a scene of strife, imperfection, and change. Observation dictated the conception of its shape. Travelers who went far north or south noticed the way the night sky changed in appearance, and by the close of the fourth century B.C., the idea of a spherical earth had obtained general acceptance in the Mediterranean area.

Rotation of the Earth

This particular view of the stars, due largely to Aristotle, persisted well into the sixteenth century. The Polish astronomer Nicolaus Copernicus then suggested that the daily motion of the celestial sphere might, after all, be an apparent one. It was more reasonable to suppose that the earth rotated beneath the sky rather than that the sky rotated round the earth. At the time (1543) the suggestion raised several arguments, one of which was that if the earth rotated, everything would be thrown off into space. Yet, as Copernicus pointed out, if the material of the earth had a strong tendency to fly apart, the celestial sphere would have a similar tendency, and to a very much greater degree in view of its immense size. Why, he asked, don't the heavens fly to pieces? But he could offer only a change in viewpoint, not a discovery; he knew of no independent ways of testing the concept of the earth's rotation.

The nature of the tests emerged in 1686 when Isaac Newton formulated his classic "laws of motion." He deduced that any two particles of matter attracted one another with a force which varied directly as their masses and inversely as the square of the distance between them. By applying this "law of universal gravitation" to the rotating earth he was able to show why objects are not hurled from its surface. The force of gravitation greatly exceeds their tendency to fly away. He also accounted for the observed motions of the moon and planets, and among other things, demonstrated that the earth's equatorial bulge was a direct result of its rotation.

In 1735 George Hadley used the concept of a rotating earth to explain the phenomenon of the trade winds. He realized that these winds were flows of air produced by heat in the tropics and deviated towards the equator by the earth's rotation. Later, in the nineteenth century, the French physicist Léon Foucault set up a swinging pendulum in the Pantheon, Paris. It underwent a slow but definite change in direction as the earth rotated and thereby provided overwhelming visual evidence to convince even the most perverse critics.

Another important feature of the night sky is the fact that different noncircumpolar stars can be seen at different times of the year. This is due to the sun's apparent motion relative to the stars. This motion takes place at a rate of about one degree per day eastward, that is, in a direction opposite to that of the stars. In one year the sun makes a full circuit of the celestial sphere, moving steadily along a path known as the ecliptic. As a result, noncircumpolar stars rise about four minutes later on successive nights, or about two hours later on successive months.

Revolution of the Earth

The Ancient Greeks thought that the motion of the sun, like that of the stars, was a real one. The sun, they declared, moved round the earth once in a year. The idea that the earth revolved in a circular orbit about the sun was first put forward by Aristarchus of Samos in the third century B.C. It remained for Copernicus, however, to work out the geometrical consequences of this in detail. Copernicus suggested that the sun's motion around the earth might be an apparent one—that if the earth traveled around the sun once in a year the sun would appear to move once round the celestial sphere in the same time. The ecliptic was not the path of the sun about the earth, but the projection of the earth's orbit on the celestial sphere.

This interpretation, like that of the daily movement of the sky, raised further difficulties, most of which could not readily be resolved. A major problem was that there was no independent way of detecting the earth's motion. Not only did astronomers think that the radius of the earth's orbit was extremely small compared with the size of the celestial sphere, but they also believed that the stars were equally remote. This being so, the revolution of the earth about the sun could not possibly produce any discernible apparent motion among the stars.

The early Christian Church adopted the Greek world-picture almost in toto and gave it a religious interpretation. It was at once a physical, moral, and spiritual universe. About its center, the earth, rotated a series of spherical shells made of the purest ether (a fifth and highly subtle fifth element) and bounded by the sphere of fixed stars. The innermost shell held the orbit of the moon and completely enclosed the earth and its atmosphere. Within this shell was the sublunar world of change, in which the interplay of the four elements, earth, air, fire, and water, gave rise to wind, rain, rainbows, meteors, comets, and other transient phenomena. Right at the center, and therefore deep in the bowels of the earth, lived Satan and his minions. Outside the moon's shell all was serene, heavenly, and eternal. The shell of the planet Venus adjoined that of the moon; then came the shells of Mercury, the sun, Mars, Jupiter, Saturn, and the fixed stars —in that order outward. Heaven lay beyond the outermost sphere. There, the great Triune God, seated on a throne placed directly above Jerusalem, bathed in celestial light, and surrounded by choirs of angels, ordered, sustained, and watched over each and every part of His Creation. It was a private universe, a universe made for man, a universe in which man represented the main aim and object of all creative activity.

Even in the sixteenth century most theolo-

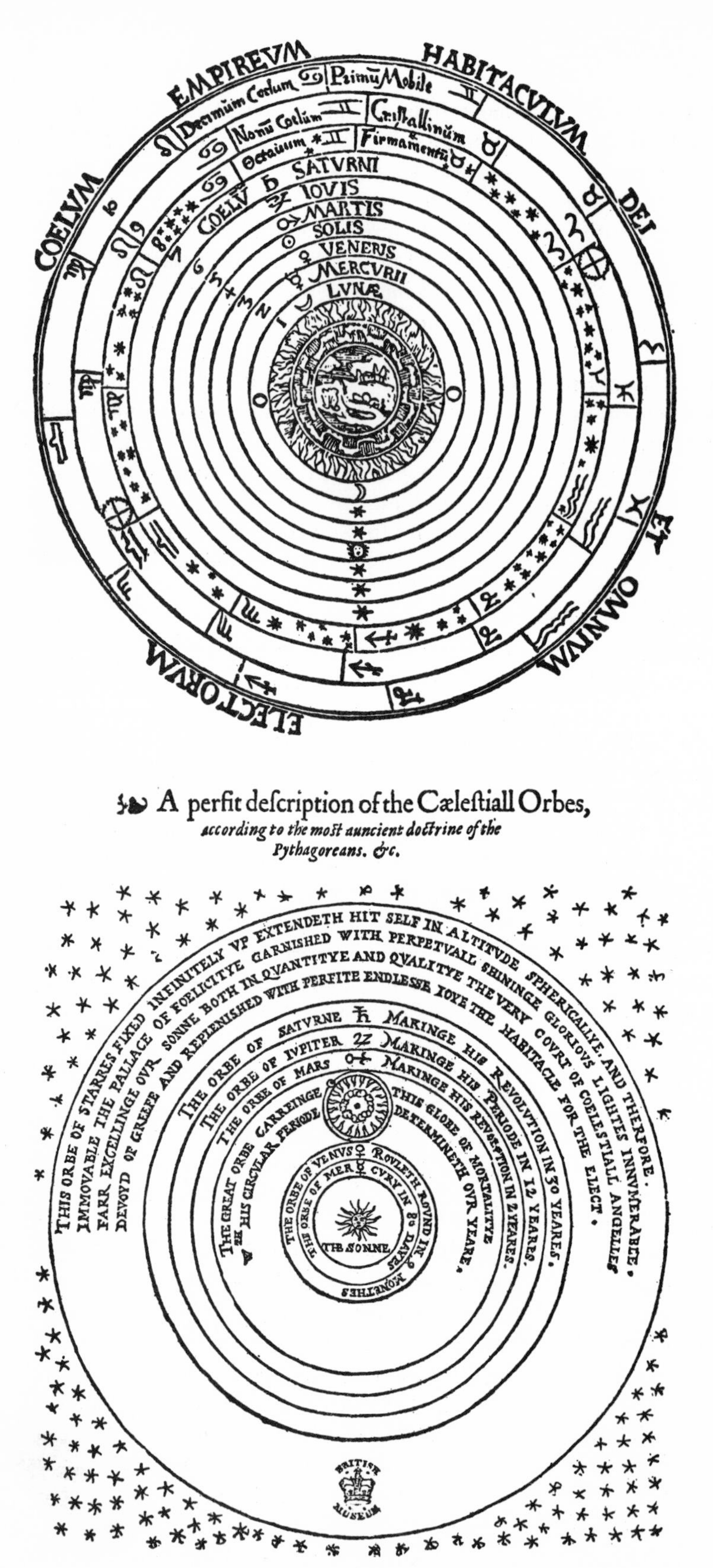

The Aristotelian universe with its fixed central earth and series of concentric shells. From Peter Apian's *Cosmographia*, 1539.

Thomas Digges's diagram of the infinite Copernican universe. From *A Perfit Description of the Caelestiall Orbes*, 1576.

gians were (not unnaturally) opposed to the idea that the heliocentric system represented the actual state of affairs. Some, however, were prepared to accept it as a working hypothesis, that is, as a convenient artifice for describing the motions of the sun, moon, and planets. That all was not well with the old geocentric world-picture became more evident in November 1572, when a bright star appeared in the constellation of Cassiopeia. Several astronomers thought that it was a comet, for comets were generally supposed to be sublunar in origin. Yet it developed no tail, rose in brilliance to rival Venus at its brightest, and most surprising of all, remained fixed in relation to neighboring stars.

By January 1573 the "new" star had fallen in brightness to equal some of the brightest stars. It then declined steadily until March 1574, when it disappeared. All attempts to detect a shift in its position were unsuccessful. Had it been sublunar it would have shown a definite parallactic effect: observers at different places would have seen it in different directions in relation to the stars. The majority of astronomers, including Tycho Brahe in Denmark, concluded that it was a starlike body generated in the celestial region well beyond the moon. This being so, the heavens could no longer be considered to be unchanging.

In 1576 the English astronomer Leonard Digges introduced the concept of an infinite Copernican universe. The sphere of fixed stars, he wrote, extended outward "in sphaericall altitude without ende." That there might be some truth in this contention became evident in 1609, when Galileo turned one of the first telescopes skyward. Although his instrument was small and imperfect by modern standards, it enabled him to see far more stars than he could ever hope to see on a clear, dark night with the unaided eye. As he surveyed the constellation of Orion with his telescope he was "overwhelmed by the vast quantity of stars." When he looked at the Pleiades he saw not only the usual six or seven stars but over forty others. And when he came to study the Milky Way he found that it was "a mass of innumerable stars planted together in clusters. Upon whatever part of it you direct the telescope straightway a vast crowd of stars presents itself to view; many of them are tolerably large and extremely bright, but the number of small ones is quite beyond determination."

Yet Galileo, a keen supporter of the heliocentric system, retained the idea of a large but finite sphere of fixed stars. Kepler, his admirer and another ardent Copernican, held similar views. In commenting on Galileo's discoveries he wrote: "And whence, pray, should we seek for conclusive evidence about the end or boundary of this visible universe, proving that it is actually a sphere of the fixed stars, and that there is nothing beyond, except from this very discovery by the telescope of this multitude of fixed stars, which is, as it were, the vaulting of the mobile universe."

When Galileo wrote: "many of them are tolerably large," he subscribed to the old belief that bright stars were bright because they were larger than faint stars. This view, based originally on naked-eye observation, seemed to be supported by appearances in the telescope. The small apertures and comparatively high magnifications of early telescopes showed bright stars as disks of light, the direct result of an optical effect known as diffraction. In passing through the finite apertures of telescopes, waves of light from stars were deflected to form small patterns instead of point images.

Sizes of the Stars In early times, opinions differed as to how large the stars actually were. Those who supported the geocentric view, Tycho Brahe among them, rejected the idea of a vast empty space between Saturn and the stars. They claimed that if such a space existed the brighter stars would not only be extremely remote but also incredibly large, perhaps larger than the orbit of the sun around the earth. They therefore put the celestial sphere just beyond the orbit of Saturn and thought that the brighter stars were about the same size as the earth. The Copernicans, on the other hand, put the stars far beyond the sphere of Saturn, and by so doing "explained" the absence of a discernible parallactic effect. In their opinion the brighter stars were as large as or even larger than the sun, and the entire solar system was no more than a point in the immensity of space.

The Discovery of Proper Motions

When the seventeenth century closed, the heliocentric system was firmly established and the sun had replaced the earth as the fundamental unit of the universe. The next important step came in 1718, when Halley announced that three bright stars (Aldebaran, Sirius, and Arcturus) had changed their positions over the course of sixteen hundred years. He compared their relative positions as given by Ptolemy, a Greco-Roman astronomer of the second century A.D., with his own observations and those of Tycho Brahe, and noted small but definite discrepancies. "What shall we say then?" he asked. "These three stars being the most conspicuous in heaven, are in all probability the nearest to the Earth; and if they have any motion of their own, it is most likely to be perceived in them."

THE SUN'S MOTION As other astronomers confirmed and extended these findings it became abundantly clear that many of the brighter stars were distributed in depth in space. This, in turn, indicated that they differed in intrinsic brightness or luminosity. The fact that two stars appeared to be equal in apparent brightness did not necessarily imply that they were equal in luminosity and distance. One star might be more luminous and more distant than the other.

The "proper motions" noticed by Halley were, of course, motions relative to the sun. It was pertinent to ask: "Is the sun itself moving relative to the nearer stars?" If so, its motion would be revealed by a general drift of the nearer stars in a contrary direction. In particular, those ahead of the sun would appear to disperse while those behind would appear to come closer together. William Herschel, among others, tried to establish whether the drift was, in fact, detectable. Unfortunately, only a few proper motions were then known, and each of these had to be resolved into two components—one arising from the motion of the star itself, and the other from the motion of the sun. This Herschel did (by using simple geometrical methods) for seven bright stars, and in 1783 concluded that the sun was traveling toward a point near the star Lambda Herculis. "We may" he wrote, "in a general way estimate that the solar motion can certainly not be less than that which the Earth has in her annual orbit." He attacked the problem again in 1805, this time with more proper motions to work with, and obtained similar results. There could be no doubt that the sun and its family of planets was moving through space.

HERSCHEL AND THE GALAXY Herschel's work showed that if, as Copernicus and Galileo had believed, the sun represented the physical, or mass, center of the universe, the center was a moving one. This seemed to be highly improbable, to say the least. Where, then, was the center? In 1785 Herschel thought that he was well on the way to finding the answer. Using reflecting telescopes 19 and 40 inches in aperture, he tried to gauge the extent or "profundity" of the realm of the stars. In doing so he became involved in the long and extremely laborious task of directing his telescope to various parts of the sky and counting the number of stars in the field of view. But he had remarkable energy and tenacity of purpose, and completed the task almost singlehanded. From the general trend of the star counts he concluded that the Milky Way was not a band or belt of stars but an optical effect. He knew that it girdled the celestial sphere and thought that its patchiness was due to variations in its depth. So, guided by the star counts, and on the assumption that the stars were more or less evenly distributed in space, he concluded that the sun lay near the center of an enormous disk-shaped system of stars. This system, "nebula," or "island universe" had an irregular outline and was deeply indented in the region of the constellation of Cygnus, where the numbers in individual star counts dropped markedly. It contained, he thought, many million stars and was one system among well over a thousand others.

The term "nebula" referred to each of a

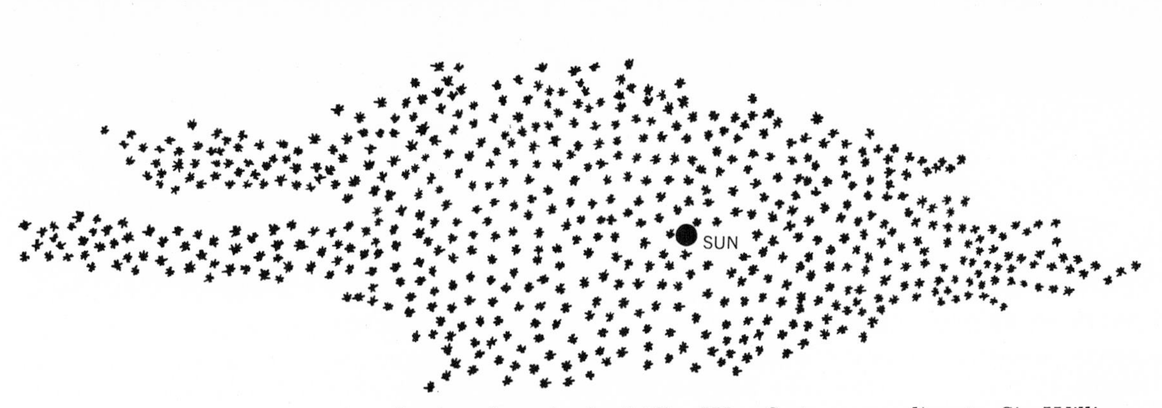

Section through the Milky Way System according to Sir William Herschel. The sun's position is indicated by the large star right of center.

number of faint, misty patches in the night sky. Between 1771 and 1784 the French astronomer C. Messier, using only a small telescope, observed and cataloged 103 objects of this kind. Herschel, with his much larger telescopes, discovered hundreds more. So many turned out to be clusters of stars that he concluded that they were all distant stellar systems independent from our own "nebula." But as his surveys progressed, so the number of nebulae increased, and he began to doubt whether they were all made up of stars. Some retained their misty appearance in even his most powerful telescopes, and he suspected that they consisted of a "shining fluid" or "self-luminous milky nebulosity." He also revised his model of the stellar system. In some regions of the Milky Way his telescopes revealed great numbers of very faint stars and, in addition, a glimmering background which defied resolution. In these regions the flattened system of stars apparently extended far beyond the space-penetrating power of his larger telescopes.

The stellar system outlined by Herschel is now called the Milky Way System, or more usually the galaxy or our galaxy. He came to realize that its stars were far from being uniformly distributed, but since the distances of even the nearest were then unknown, he had no idea of its size.

First Determination of Stellar Parallax

The first reliable measurement of the parallax of a fixed star was made in 1838 by the German astronomer F. W. Bessel. It concerned 61 Cygni, a star whose comparatively large proper motion of 5.2 seconds of arc a year indicated that it was one of the sun's nearer neighbors. He used an ordinary trigonometrical method

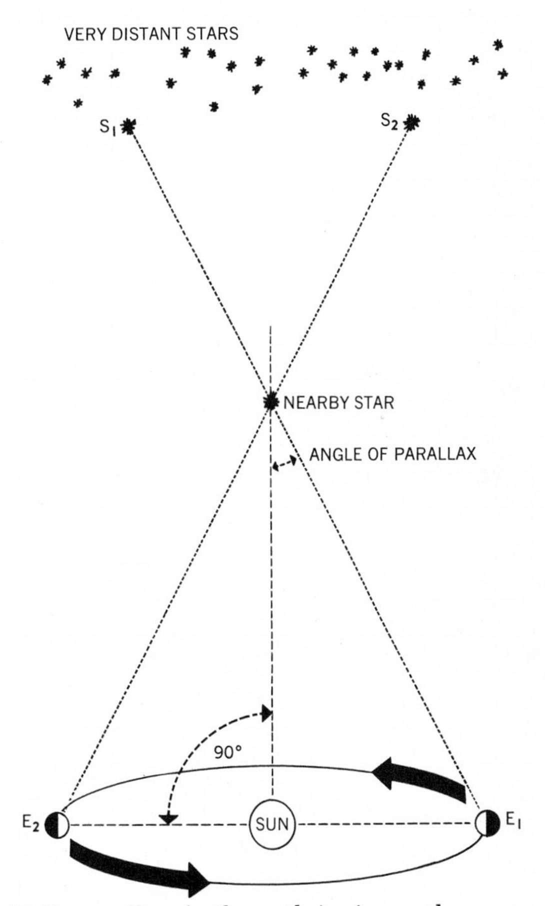

Stellar parallax. As the earth in six months moves from E_1 to E_2, a nearby star seems to move from S_1 to S_2. The angle of parallax is the angle subtended by the radius of the earth's orbit as seen from the star. The size of the earth's orbit is greatly exaggerated.

similar to that used in surveying. But, where a surveyor might take a base line of a few miles to determine the distance of an inaccessible mountain, Bessel took as his base line the diameter of the earth's orbit. The observations consisted of measuring the apparent distance of 61 Cygni from two neighboring but faint stars at intervals of six months. The faint stars, assumed to be so distant as to have no perceptible parallax, acted as fixed reference points. Much to Bessel's delight 61 Cygni did change its relative position, and by an amount which put its distance at 600,000 astronomical units, where 1 astronomical unit (1 A.U.) is nearly 93 million miles, or the average distance of the earth from the sun. More accurate determinations of the angle of parallax have since set the distance at 701,300 A.U.

Since star distances expressed in astronomical units yield very large numbers, astronomers now prefer to work with more convenient units. One of these, the parsec, is the distance which corresponds to a parallax of one second of arc. Another unit, frequently used in popular writing, is the light-year. This is the distance which light travels in a year. Light has a velocity of 186,282 miles a second, so in a year it travels about 6 million million miles. One parsec is equal to 3.26 light-years, so when we use these units the distance of 61 Cygni becomes 3.41 parsecs, or 11.1 light years.

Technical Advances in the Nineteenth Century

Developments in astronomy in the nineteenth century showed that our galaxy was much more complex than Herschel had thought. One of these involved the use of the spectroscope, a device for analyzing starlight. It not only enabled astronomers to study the chemistry and physics of the stars but also revealed that some nebulae, like the Great Nebula in Orion, were gaseous, while others, like the Great Nebula in Andromeda, consisted of stars. It also made possible a determination of a star's radial velocity, or its velocity in the line of sight. This, considered in conjunction with the proper motion, led directly to a knowledge of the actual speed and direction of motion of the star in space. Through another development, photography, astronomers could record large areas of the night sky and study at their leisure the complex form and structure of the Milky Way and the general distribution of stars, star clusters, and nebulae. One important outcome of this was the discovery that the Milky Way owed its patchy appearance largely to the presence of great clouds of obscuring interstellar dust. Another was the extension of the method of trigonometrical parallaxes to fainter and therefore (in the main) more distant stars.

Size of the Galaxy

At the end of the nineteenth century astronomers realized that Herschel had greatly underestimated the size of the galaxy. Some thought that the galaxy composed the entire universe and so contained all the several thousand known nebulae. Others, impressed by the fact that apparently small and faint nebulae tended to avoid the region of the Milky Way, argued that the galaxy was just one of many similar but possibly smaller systems. Estimates of its diameter ranged from about 18,000 to 40,000 light-years, but these were little more than guesses. Almost everyone, however, agreed that the sun was at or near the center. This, they thought, was certain.

The early years of the present century saw great advances on both the observational and theoretical fronts. One observation disposed of the old idea that the stellar system was a chaos of stars, each with its own independent motion. From a statistical study of a large number of proper motions, J. C. Kapteyn of Gröningen, Germany, found that the stars tended to drift or stream in two definite directions relative to the sun. When allowance was made for the effect of the sun's motion, the stars concerned formed two intermingled stellar swarms moving in opposite directions in space. The two directions lay in the plane of the Milky Way and were clearly related to some fundamental dynamical property of the system as a whole. In 1926 B. Lindblad explained what that property was: the galaxy rotated, not in the same way as a wheel, but with an angular velocity that varied with distance from its center.

Shapley's Discoveries

In another line of attack, the American astronomer H. Shapley took advantage of a remarkable relationship between the luminosities of certain variable stars and the periods of their light variations. With stars of this kind, known as Cepheids, one had only to measure the period of light variation to establish the luminosity. Once this was known it could be compared with the apparent brightness to determine distance. The connecting link, as the reader has probably surmised, was the inverse-square law of light, or the fact that the brightness of a star decreases as the square of its distance increases.

Fortunately Cepheids are highly luminous and can be seen to immense distances—hence their special value as distance indicators. Shapley used them to determine the individual distances of remote objects known as globular star clusters. Although each of these had a decidedly nebulous appearance in small telescopes, large instruments, assisted by photography, resolved it into a ball-like swarm of many thousands of stars. Shapley found that the clusters lay just outside the main body of the stars of the galaxy and thus were valuable guides to its size. The nearest, an object named Omega Centauri, was about 20,000 light-years away, while the most distant was about 230,000 light-years away.

Shapley then studied the distribution of the one-hundred-odd globular clusters in space. They formed, he found, a great but roughly defined ellipsoidal system, symmetrically divided by the plane of the Milky Way. The center of the system lay somewhere in the direction of Sagittarius, where the star clouds of the Milky Way were particularly bright. He concluded that the center of the galaxy not only lay in this direction but also coincided with the center of the system of globular clusters. The geometrical features of the latter then indicated that the sun lay about 60,000 light-years from the center of a highly flattened

The globular star cluster Omega Centauri. It contains many thousand stars, but owing to its great distance looks like a faint hazy star to the unaided eye.
—Radcliffe Observatory, Pretoria

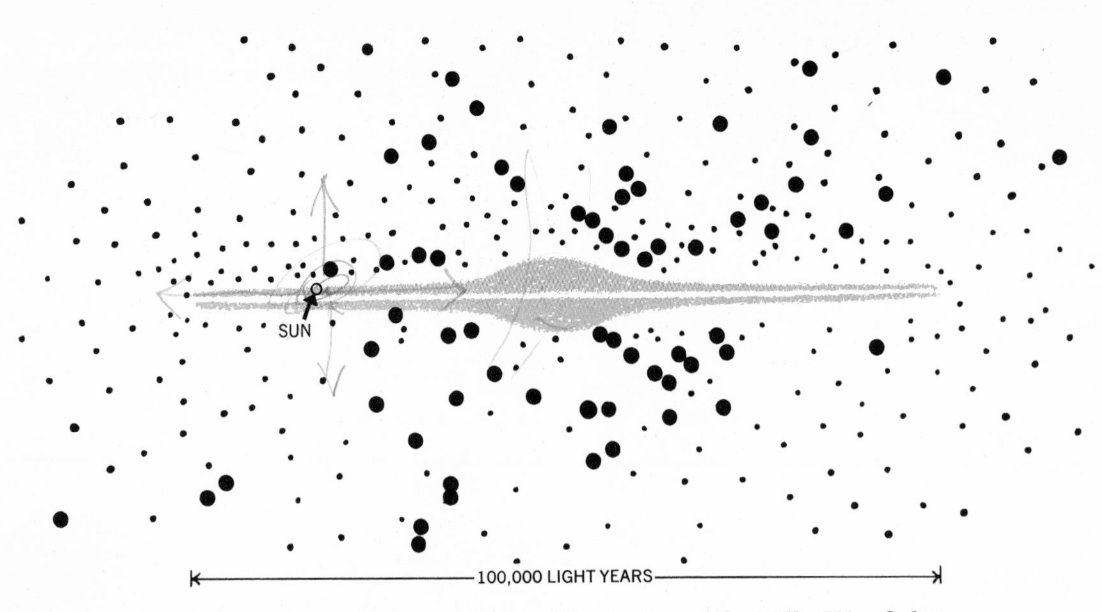

Cross-section of the Milky Way Galaxy

aggregation of stars at least 300,000 light-years in diameter.

Shapley showed beyond all reasonable doubt that the galaxy, despite its vast extent, was not the universe. Using Cepheids as distance indicators, he found that the Clouds of Magellan, two prominent star clouds in southern skies, were so far outside the Milky Way System as to rank as galaxies in their own right. The same was true of the Great Nebula in Andromeda, the thirty-first entry in Messier's catalog, and for this reason, commonly designated M31. Even so, M31, much larger than either of the Clouds of Magellan and at an estimated distance of 750,000 light-years, was still no more than a tiny island compared with the continent of the galaxy. Once M31 achieved extragalactic status it was logical to assume that all other apparently smaller spiral nebulae were extragalactic. If that were so, and if apparent size were taken as a rough guide to distance, they could be anything from 10 million to 100 million light-years away.

Rotation of the Galaxy

In a general way the modern picture of the universe still conforms to that presented by Shapley in 1918. One early modification, made by Shapley himself, was to reduce the diameter of the galaxy to 180,000 light-years. Researches by others had revealed the existence of a widespread galactic haze produced by interstellar dust. This both dimmed and reddened the light of very distant stars and globular star clusters, making them appear more distant than they really were. But even this reduction was not enough, and the diameter is now reckoned to be about 100,000 light-years. Dynamical considerations support a figure of this order.

Lindblad's suggestion that the galaxy rotated was taken up by J. Oort of Leiden. He confirmed Shapley's surmise that the galactic center lay in the direction of the star clouds of Sagittarius and showed that, in general, the orbital velocity decreased with increasing distance from the center. The sun's orbital speed is now reckoned to be about 160 miles a second, while its distance from the center is thought to be 33,000 light-years. This gives the sun an orbital period of roughly 250 million years and makes the mass of the system equal to at least 100 billion suns.

Contents of the Galaxy

The galaxy, then, is a rotating system of stars, gas, and dust so enormous that the solar system, although immense in size by earthly standards, is a mere speck in comparison. Some idea of the variety and nature of its contents will emerge in the following pages. We shall see, among other things, that it has a spiral structure, first detected through optical studies of the distribu-

Star cloud in Sagittarius, the direction of the center of the galaxy.
—ROYAL ASTRONOMICAL SOCIETY

tion of hot, young stars, and more recently revealed in much greater detail and extent by radio observations of interstellar clouds of neutral hydrogen. We shall also see how its stars, born from gas and dust, form a vast population made up of at least two subpopulations and probably containing stars of several different generations. The modern galaxy is essentially an evolving system in the sense that it is going through an individual life history, the general trend of which we are only just beginning to discern. But this is also true of other galaxies. Some 4 billion of them are believed to lie within the reach of the 200-inch telescope of the Mount Wilson and Palomar Observatories. They have different shapes and sizes, and judging by their contents, are of different ages. They also have velocities of recession which increase with increasing distance from our own galaxy. About 30 percent of them have a spiral structure, and of these M31 appears to be one of the largest.

A Universe of Galaxies

M31, at an estimated distance of 2.2 million light-years, has a diameter of about 180,000 light-years—certainly larger than our galaxy's. Yet it is completely dwarfed in size and mass by giant spheroidal galaxies like M87 in Virgo, and these in turn are dwarfed in overall energy output by the recently discovered quasi-stellar

M31, the Great Spiral Galaxy in Andromeda
—MOUNT WILSON AND PALOMAR OBSERVATORIES

radio sources, one of which is at an estimated distance from us on the order of 10 billion light-years. This particular object, designated 3C–9 and referred to in our final chapter, is thought to lie at the extreme outer limit of the observed universe.

If we reduce the diameter of the solar system to 1/10 inch or the size of a large pinhead, Proxima Centauri, the nearest star after the sun, will be a submicroscopic speck 28 feet away, and the galaxy, a flattened disk about 140 miles in diameter. If we now reduce the galaxy to a pinhead, all the galaxies so far observed will be contained in a sphere of 500-foot radius. Models of this kind, though they emphasize the almost overwhelming immensity of space and the corresponding smallness of the earth, are misleading in many respects. In particular, we must not think that our galaxy lies at the center of the universe of galaxies. It *appears* to be at the center, but as we shall see later, the universe of galaxies has no absolute center. There are as many centers as there are galaxies!

In fact, the naked-eye view of the night sky bears no resemblance whatever to the modern astronomical universe. Only about 2,500 stars can be seen on a clear, moonless night, and the Milky Way presents its age-old appearance of a faint, irregular band of light. The entire scene turns around an earth which, in the light of everyday experience, seems to be flat and fixed. Small wonder then that man, for the greater part of his history, thought that he lived at the center of a "private" physical universe, the parts of which traveled around him, existed for his own special benefit, and like his own spiritual nature, were completely indestructible.

CHAPTER II

MAPPING THE STARS

It has been the custom from early times to divide the stars into groups which received the names of familiar objects, or more often, of persons and animals famous in mythology. These constellation figures, as they are called, are purely imaginative things. In a few instances (for example, Scorpius, the Scorpion, and Orion, the Hunter) a particular star group does suggest a particular figure, but in most cases the groupings bear no resemblance to the shapes they are supposed to represent.

Constellation Figures

Different peoples have formed their own sets of constellation figures. In ancient times those of the Egyptians bore no resemblance to those of the Chinese, and both differed from those of the Greeks. More recently the Polynesians, Maori, and others have devised further sets of figures, but these have few if any points of resemblance with one another. Though the practice was probably introduced for reasons of convenience, to provide a rough but useful way of describing the night sky, it was also an expression of man's primitive urge to animate nature—to project his childlike fancies on inanimate objects in the world around him. By peopling the sky with figures and weaving stories about them, he brought it into closer relationship with himself and his affairs. It became, to his way of thinking, a moving, living picture book, and therefore a scene of much greater interest than before.

THE ZODIAC The earliest constellation figures were probably the twelve in the region of the zodiac, that part of the sky in which move the sun, moon, and five naked-eye planets. The zodiac takes the form of a band or belt about 15 degrees wide, centered on the ecliptic. The fact that the sun, moon, and planets (with the exception of Pluto) stay within its limits is due solely to the fact that the solar system is rather flat in shape. The planes of the orbits of the moon and planets all lie close to the plane of the earth's orbit.

Some historians think that the zodiacal apellations had their origin with either the Sumerians or the Babylonians, but the evidence for this is somewhat slender. Even so, the need to divide the zodiac into a convenient number of equal divisions must have arisen at a very early period. In Greek times the divisions were: Aries, Taurus, Gemini, Cancer, Leo, Virgo, Chelae, Scorpius, Sagittarius, Capricornus, Aquarius, and Pisces. Chelae, the Claws of the Scorpion, was changed to Libra, the Scales, sometime during the first century B.C.

Precession of the Equinoxes

Over the centuries the zodiac acquired great significance in astrology and not everyone distinguished between the so-called signs and the corresponding constellations. The former refer to the names or symbols given to the twelve divisions. Each 30 degrees in extent, these divisions begin at the vernal equinox, or first point of Aries. At this point the sun, traveling along the ecliptic in a gradually increasing northerly direction, crosses the celestial equator. The equinox, however, is a *moving* point, traveling backward in relation to the stars at a rate of about 50.26 seconds of arc a year. The same is also true of the autumnal equinox, for the celestial equator, although inclined at a more or less constant angle to the great circle of the ecliptic, moves bodily around the latter once in 25,800 years. This motion is referred to as the precession of the equinoxes. It means that the vernal equinox fell in the constellation of Aries about two thousand years ago and in Taurus some four thousand years ago. At present it is moving through the constellation of

Pisces toward that of Aquarius. The signs of the zodiac have therefore drifted away from their corresponding constellations, a feature which astrologers conveniently ignore for their own purposes.

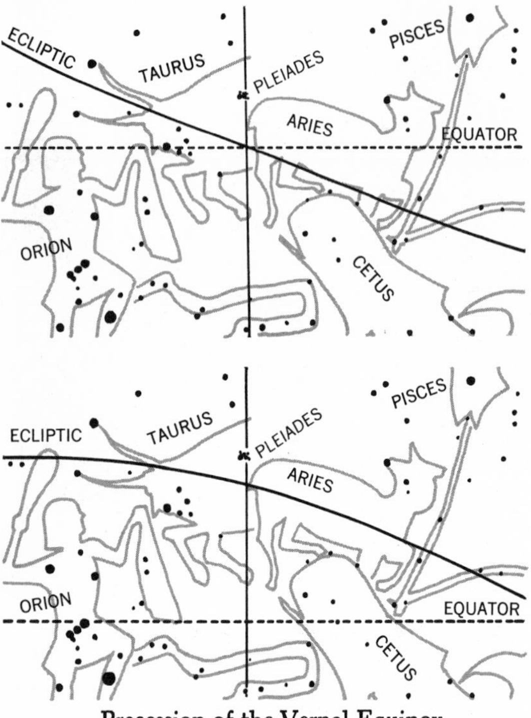

Precession of the Vernal Equinox.

Constellations and Star Names

Most of the present-day constellations were identified by the Greeks. Some are referred to by Homer and date to at least the ninth century B.C. He refers to Boötes, the Wagoner, and writes that the Great Bear "is also called by the name of the Wagon." The latter "turns about heavens' axle tree" (the north celestial pole) and "holds ope a constant eye upon Orion." He also knew about circumpolar stars, for he observes that while some constellations go below the skyline, the Wagon "alone is without lot in the bath of the ocean." One or two of the brighter stars had already been named by this time. Both Homer and Hesiod refer to the late summer rising of Sirius (called Sothis by the Egyptians), the brightest star in the night sky, while Hesiod mentions Arcturus and the two prominent star clusters Pleiades and Hyades.

In the third century B.C. Aratus wrote the *Phenomena*, a poem of great merit and charm, which described forty-five constellations, all but three of which tally with those of today. His lines about the Pleiades are particularly fine:

> *Near his [Perseus] left knee the Pleiads next are roll'd*
> *Like seven pure brilliants set in a ring of gold.*
> *Though each one small, their splendour all combine*
> *To form one gem, and gloriously they shine.*
> *Their number seven, though some men fondly say,*
> *And Poets feign, that one has pass'd away.*
> *Alcyone—Celoeno—Merope—*
> *Electra—Taygeta—Sterope—*
> *With Maia—honor'd sisterhood—by Jove*
> *To rule the seasons plac'd in heaven above.*
> *Men mark them, rising with the solar ray,*
> *The harbingers of summer's brighter day—*
> *Men mark them, rising with Sol's setting light,*
> *Forerunners of the winter's gloomy night.*
> *They guide the ploughman to the mellow land—*
> *The sower casts his seed at their command.*

An Early Star Globe

According to tradition, Aratus obtained much of his information from a celestial globe made by Eudoxus a century earlier and placed on exhibition in Athens. Cicero writes that "a very learned man" told him that Eudoxus adorned his globe "with the fixed stars of heaven, and with every ornament and embellishment." Eudoxus also appears to have had some connection with the earliest extant globe, for the arrangement of the forty-two figures cut on its surface suggests that it was probably constructed around 300 B.C. Among the "missing" figures are Ursa Major, Ursa Minor, and of course, those devised much later for stars near the south celestial pole. The globe, 26 inches

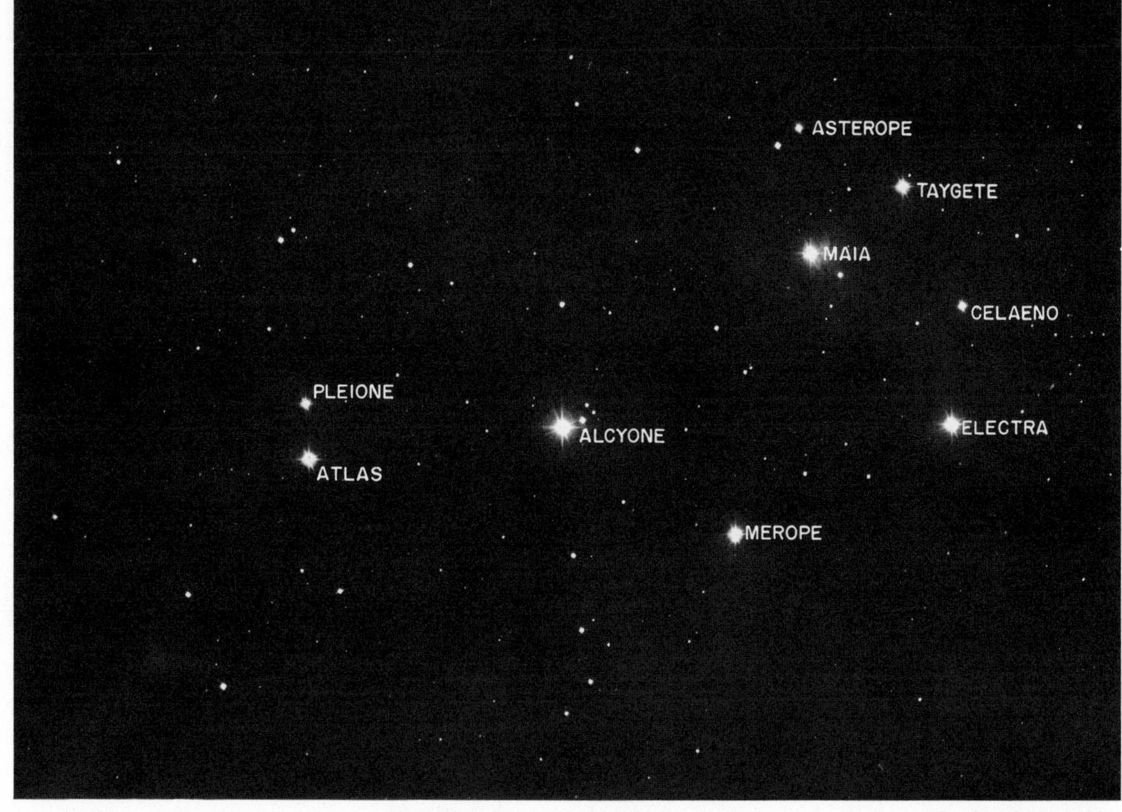

The Pleiades, a prominent naked-eye star cluster. The brightest star, Alcyone, is of the third magnitude.
—Royal Observatory, Edinburgh

in diameter and made of marble, is supported on the broad shoulders of a statue of a kneeling Atlas some six feet high. The entire sculpture is in the National Museum of Naples and is known as the Atlante Farnesiano.

The First Star Catalog

The first catalog of stars was prepared by Hipparchus, the greatest Greek astronomer of the pre-Christian era. In 136 B.C. he observed a bright, starlike comet in the constellation of Scorpius, and to plot its course and those of any future visitors, began to measure the relative positions of the brighter stars. For this task he used circles graduated into degrees and parts of a degree of angle and fitted with naked-eye sights. The outcome was a catalog of the positions of 1,080 stars contained in forty-eight "asterisms," or constellations. Of these twenty-one were northern, fifteen were southern, and the remainder occupied the zodiac. The star positions were referred to the ecliptic, that is, they were given in terms of celestial longitude (measured eastward from the vernal equinox) and celestial latitude (measured north and south of the ecliptic). This was a significant advance, for it meant that a star's position could be uniquely specified on a coordinate system instead of on the arbitrary shape of a constellation figure.

Discovery of the Precession of the Equinoxes

The value of the new method soon became apparent, for when Hipparchus compared certain star positions with those recorded by Greek observers some 150 years earlier, he noticed that the bright star Spica had apparently moved nearer the autumnal equinox by about 2 degrees. Other stars, he found, had also shared in this slow west-to-east drift, so it was pretty clear that the equinoxes were precessing. The discovery was probably made without knowledge of the earlier work of Kidinnu, a Babylonian astronomer, who according to some historians of science, discovered the precession of the equinoxes sometime during the third century B.C.

Since the ecliptic is a great circle and therefore has two poles on the celestial sphere, the effects of precession are not restricted to the motion of the equinoxes. As the latter travel backward along the celestial equator at the rate of one revolution in 25,800 years, each celestial pole moves round the corresponding pole of the ecliptic at the same rate. This means that different stars can take over the role of North Pole star. At the time of Hipparchus, Polaris was about 12 degrees away from the celestial pole. Since then it has moved closer to the pole and will be at its closest (about 28 minutes of arc) in the year 2102. Six thousand years ago the pole star was Thuban Draconis. In twelve thousand years' time it will be Vega, a bright star at present about 50 degrees away from the pole.

Star Magnitudes

Hipparchus divided the stars into six classes according to their brightness. The brightest were of the first "size" or "magnitude"; those a little less bright were of the second magnitude; those a little less bright again were of the third magnitude, and so on to those of the sixth, the faintest visible to the unaided eye. This system formed the basis of all subsequent work in this field. A star of the first magnitude is now reckoned to be one hundred times brighter than one of the sixth magnitude. It

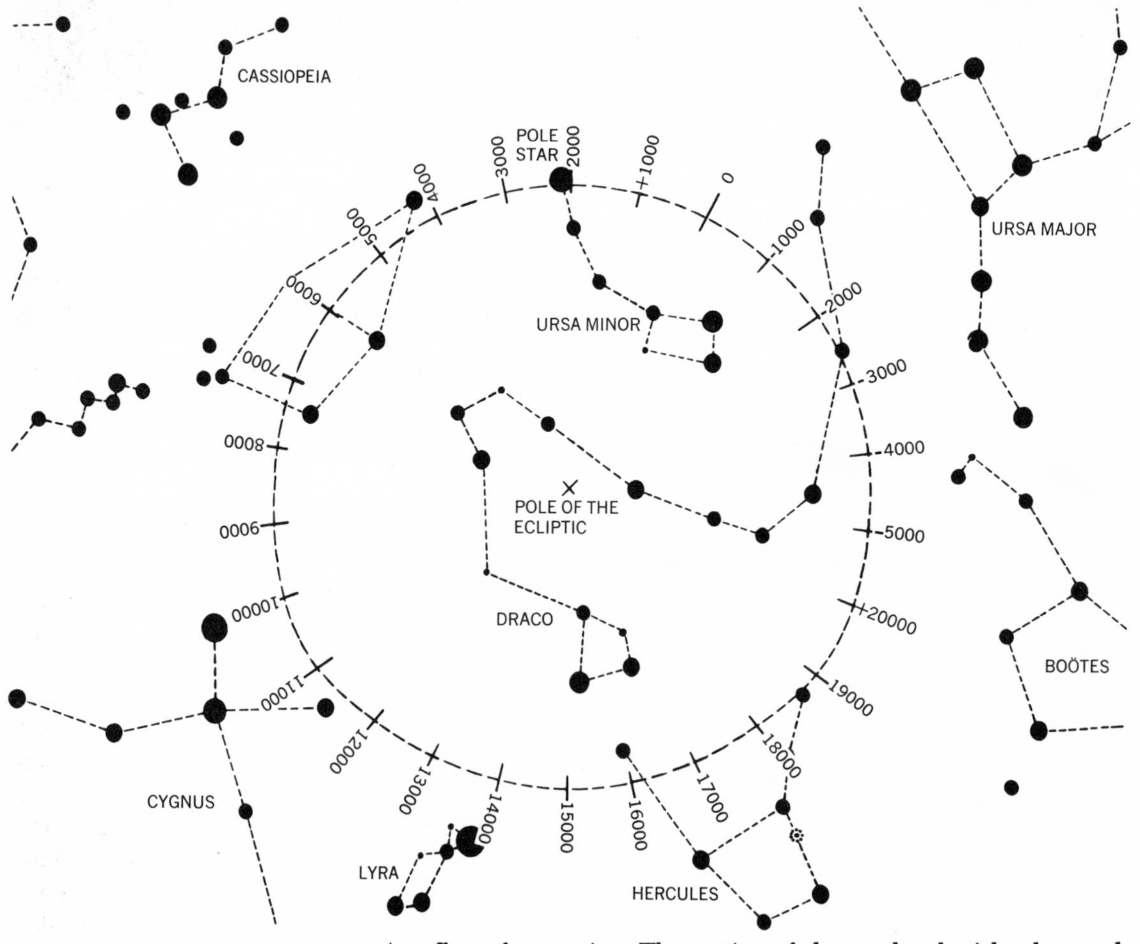

An effect of precession. The motion of the north celestial pole round the north pole of the ecliptic.

follows that a difference of one in magnitude corresponds to a brightness ratio of the fifth root of 100, or 2.512. A star of the first magnitude is 2.512 times brighter than one of the second, one of the second is 2.512 times brighter than one of the third, and so on in regular progression.

Unfortunately for astronomers, the stars do not fall conveniently into six classes of brightness. The telescope reveals stars much fainter than those of the sixth magnitude, stars broadly assigned to a particular magnitude class are not necessarily equal in brightness, and there are objects brighter than stars of the first magnitude. The modern magnitude scale is therefore both extensive, well subdivided, and at one end proceeds through zero to negative numbers. Its limits are determined by the brightest star (the sun, magnitude −26.7) and the faintest objects photographed with the 200-inch Hale telescope (very remote galaxies, magnitude 23). The difference in magnitude between these two extremes is therefore about 50, and this corresponds to a brightness ratio of 100 billion billion.

Sirius is of magnitude −1.42. The difference in magnitude between the sun and Sirius is therefore about 25, which corresponds to a brightness ratio of 10 billion. This means that it would require 10 billion stars as bright as Sirius to replace the sun in brightness. Similarly, Sirius is roughly 10 billion times brighter than the faintest objects photographed with the 200-inch telescope.

The work of Hipparchus is known largely through the writings of Ptolemy, the last great astronomer of antiquity. Ptolemy systematized the bulk of Greek contributions to astronomy and thus ensured its safe but slow transmission through the Muslim world to Western Europe.

Greek and Arabic Star Names

The Muslims, for their part, tended to copy and embellish Greek findings. In some areas they made notable additions, one of which was to give individual names to a large number of stars. The Greeks and Romans knew only a few stars by name. Aratus, for instance, named the seven brightest stars of the Pleiades and referred to Arctophylax (Arcturus), Aix (Capella), Findemiator (Epsilon Virginis), Prokuon (Procyon), and Seirios (Sirius). Eratosthenes mentioned Stachys (Spica) and also Canopus, a star just visible from Rhodes but not from Athens. To these Ptolemy, or perhaps Hipparchus, added Regulus, the Little King, and Antares, the Rival of Ares (Mars). Latin names like Bellatrix, Castor, Pollux and Polaris are of a later but uncertain date. Some of these names the Muslims rendered into Arabic either by translation or rough transcription. They then devised other names for practically all the brighter stars down to and including those of the third magnitude. In many cases the name of a star described its position in a constellation. Thus Fum-al-haut (now written Fomalhaut) is the Mouth of the Southern Fish; Al-ghul (now Algol) is the Demon; Ridjil-al-djauza (Rigel) is the Leg of the Giant.

A more convenient nomenclature denotes the brighter stars in each constellation by letters of the Greek alphabet. Thus Betelgeuse, the brightest star in Orion, is also called Alpha Orionis. Rigel, the second brightest, is Beta Orionis, Bellatrix, the third, is Gamma Orionis, and so on. This method was introduced in 1603 by J. Bayer of Augsburg, but in some cases his arrangement no longer holds good. For instance, Pollux, or Beta Geminorum, is brighter than Castor, or Alpha Geminorum. This does not necessarily mean that these stars have changed in comparative brightness since Bayer's time. He sometimes based his sequences not on brightness but on the positions of the stars in a figure. In any case, well over one half of his magnitudes were based on the somewhat unreliable brightness estimates of Ptolemy and Tycho Brahe.

An Early Star Atlas

Bayer prepared an atlas, *Uranometria,* in which he depicted sixty constellation figures and cataloged 1,300 naked-eye stars. Forty-eight of the constellations were based on the classical figures described by Aratus. They resembled in many respects the excellent woodcuts of the constella-

Celestial Map, Northern Hemisphere, from a woodcut by A. Dürer.
—Rosenwald Collection, National Gallery of Art, Wash., D.C.

tion figures made in 1515 by the great German artist, Albrecht Dürer. The other twelve, added by Bayer himself, were Apus, Toucana, Grus, Phoenix, Dorado, Pisces Volans, Hydrus, Chamaeleon, Musca Australis, Triangulum Australis, Indus, and Pavo. These belong to the Southern Hemisphere and were based on information given by navigators who had sailed well south of the equator. Since Bayer's time a further twenty-eight figures have been introduced, most of them in the Southern Hemisphere, thereby making eighty-eight altogether. The figures, however, have become little more than historical curiosities.

Right Ascension and Declination

Although every star has a place in a constellation, its position on the celestial sphere must be expressed in precise terms. This is done by using a coordinate framework based on the celestial equator and having two coordinates—right ascension (R.A.) and declination (Decl.).

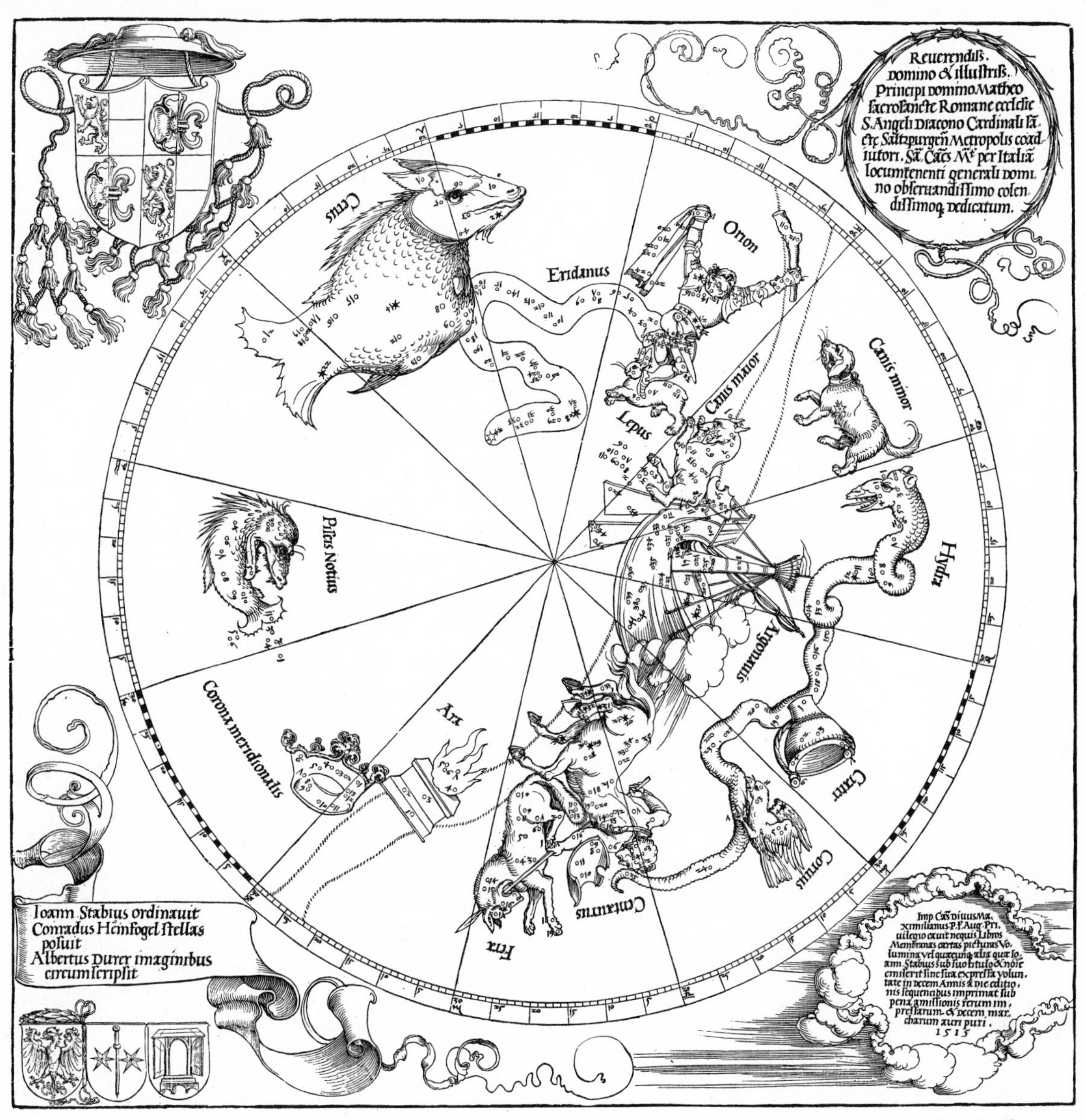

Celestial Map, Southern Hemisphere, from a woodcut by A. Dürer. —ROSENWALD COLLECTION, NATIONAL GALLERY OF ART, WASH., D.C.

The latter are analogous to longitude and latitude on the earth. Right ascension corresponds to longitude and is measured eastward from the first point of Aries. It can be expressed in angular measure but is usually given in units of time: 15 degrees corresponds to 1 hour, 1 degree to 4 minutes, and so on. Declination corresponds to latitude and is reckoned positive for stars north of the celestial equator and negative for those south. The main disadvantage of this system is that the steady precession of the equinoxes brings about a slow increase in the coordinates of the stars. Modern charts and catalogs of stars are therefore based on the position of the first point of Aries for 1950.0. Corrections can then be applied if the coordinates for some other time or epoch are required.

Atmospheric Refraction

The merit of recording star positions in terms of right ascension and declination was recognized toward the end of the sixteenth century by Tycho Brahe. Using large sextants, quad-

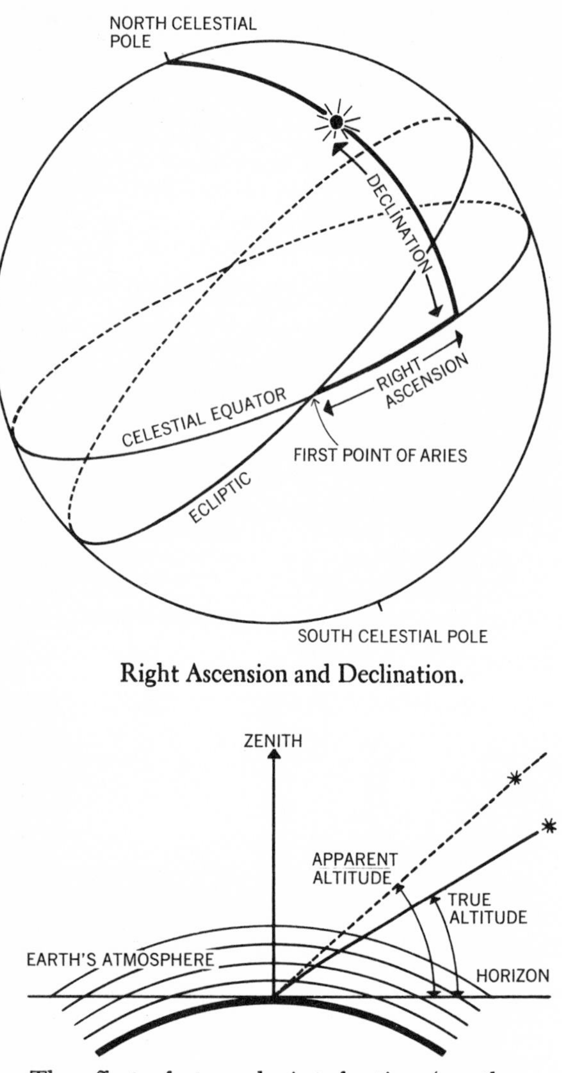

Right Ascension and Declination.

The effect of atmospheric refraction (greatly exaggerated).

rants, and armillaries, all fitted with naked-eye sights, Tycho determined the positions of the brighter stars and followed the movements of the sun, moon, and planets over many years. At an early stage in this work he investigated the effects of atmospheric refraction. This phenomenon, as the name implies, is the slight change in direction which light undergoes when it passes through the earth's atmosphere. As a result, stars and other distant bodies appear to be displaced upward by amounts which depend on their altitudes. At the horizon the displacement is about ½ degree, or roughly the apparent diameter of the sun or moon. It then decreases rapidly with increasing altitude to become 58 seconds of arc at 45 degrees and zero at 90 degrees, the zenith. One of Tycho's many achievements was to determine the amount of refraction from observations and then to apply it as a correction to his results.

Early Star Catalogs

Tycho prepared a catalog of the positions and magnitudes of 777 stars for the epoch 1600.0. He also constructed a brass globe, almost six feet in diameter, which carried the positions of 1,000 stars. "A globe of this size," he wrote, "so solidly and finely worked, and correct in every respect, has never I think been constructed up to now . . . anywhere in the world. (May I be forgiven if I boast.)" Yet J. Flamsteed, less than a century later, found that Tycho's star positions were generally uncertain to 3 or 4 minutes of arc and "often erred ten minutes or more."

While first Astronomer Royal at the Royal Greenwich Observatory, Flamsteed used instruments fitted with telescopic sights and spent over thirty-three years preparing a better catalog. His collected results, the *Historia Coelestis,* first published in 1712, contained, among other things, the coordinates of nearly 3,000 stars. Their general uncertainty was about 10 seconds of arc, a great achievement at the time but an order of accuracy which no modern astronomer could possibly tolerate. Flamsteed also introduced the expedient of numbering the naked-eye stars in a constellation, not in order of brightness, but in order of position, from west to east. In doing so he included all the bright stars in Bayer's catalog, but these are now referred to by their Greek letters in preference to their Flamsteed numbers.

Halley, Flamsteed's immediate successor at Greenwich, found that some stars had changed their positions in relation to other stars. This was the discovery of the proper motions of stars, which showed, among other things, that star catalogs had to be prepared with the utmost possible accuracy. Unless this were done, later astronomers would be unable to tell whether an apparent shift in a star's position was due to its proper motion or to an error in the comparison catalog.

Bradley's Discoveries

J. Bradley, the third Astronomer Royal, made an equally surprising discovery. In 1729, before he had taken up duties at Greenwich, he announced that every star undergoes an apparent displacement resulting from an effect known as the aberration of light. He discovered the displacement while trying to detect stellar parallax. He found that in the course of a year the star Gamma Draconis described a small ellipse. But although this behavior was of the type which he had anticipated, the position of the star in its ellipse was always 90 degrees from its expected parallax position. In other words, the displacement of this and other stars was always toward a point on the ecliptic 90 degrees behind the sun. Further, the amount of the displacement (about 20.5 seconds of arc at most) depended on the sine of the angle between the direction of the earth's motion and the direction of the star. This led Bradley to conclude that the light from a star undergoes an apparent shift in direction arising from the combined effect of our own velocity (due to the earth's orbital motion of roughly 18.5 miles a second) and the velocity of light. A similar effect is observed when driving an automobile through a shower of rain. Although the rain may be falling vertically it appears to be moving toward the windshield because of the combined effect of its velocity and that of the automobile.

Bradley found that the effects of precession and the aberration of light could not completely account for the observed displacements of Gamma Draconis. There remained a residual displacement of a type which indicated that the north celestial pole moved slightly toward and away from the north pole of the ecliptic in a period of nearly nineteen years. This nodding, or nutation, as it is called, means that the earth's axis has a slight wobble in addition to its precessional motion.

Bradley's discovery of the aberration of light supported the Copernican doctrine of the earth's orbital motion. It also offered a way of demonstrating the earth's rotation. Our velocity relative to the sun has two components: one arises from the earth's orbital motion and the other from its rotation. The latter, like the orbital motion, also produces an apparent displacement of the stars. Its amount is extremely small, being only 0.32 seconds of arc at most even for an observer at the equator, and its detection lay well outside the scope of Bradley's instruments.

While at Greenwich, between 1742 and 1762, Bradley made a vast number of observations with instruments of improved accuracy and performance. Later, Bessel of Königsberg studied them in great detail and in 1818 issued a catalog of three thousand stars. The positions, based on Bradley's measurements and referring to the epoch 1760.0, had an average error of less than 4 seconds of arc in declination and 15 seconds of arc in right ascension.

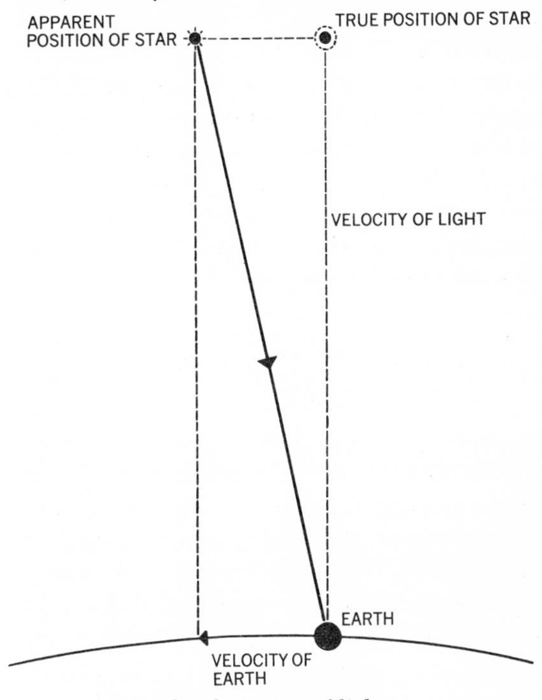

The aberration of light.

The Bonn Durchmusterung

Although the Bessel-Bradley catalog proved invaluable in various aspects of practical astronomy, many astronomers required a catalog that was much more comprehensive. This need was met between 1859 and 1862 by F. W. A. Argelander, Director of Bonn Observatory. With the help of a few able assistants he cataloged all the stars down to the ninth or tenth magnitude between declinations +90 and −2 degrees. E. Schönfeld, his chief assistant and successor at Bonn, then extended the catalog to declination −23 degrees. The combined work, known as the *Bonn Durchmusterung*, contained entries of the positions and magnitudes of a total of 458,000 stars and was accompanied by charts which greatly simplified the identification of an observed star. It formed the basis of studies of stellar distribution and is still highly valued by astronomers, but more because of its extensive nature than for its accuracy. A similar catalog, again with charts, was later prepared at Cordoba, Argentina. It reached to declination −90 degrees and gave the positions and magnitudes of about 614,000 stars.

Recording Star Positions by Photography

In September 1882, a bright comet appeared and D. Gill, H. M. Astronomer at the Cape Observatory, South Africa, arranged for it to be photographed with an ordinary camera strapped to a telescope. Although astronomical photography was then in its infancy, Gill obtained some remarkably good pictures. What impressed him most was the great number of stars that appeared along with the comet. He realized that he had hit on a much less laborious way of determining the positions and magnitudes of large numbers of stars and lost no time in making plans for a photographic chart and catalog of southern stars. The photography took five years (1885–1890). J. C. Kapteyn then undertook the immense task of measurement and reduction of nearly half a million star images, and the completed work, the *Cape Photographic Durchmusterung*, was published in 1900.

Meanwhile other astronomers had become aware of the great possibilities of photography for recording star positions. At an international congress in 1887 it was decided to prepare a photographic atlas of the entire sky, together with a catalog of the positions and photographic magnitudes of all stars down to about magnitude 14. This great project, the *Astrographic Chart and Catalogue*, originally involved eighteen observatories, but all manner of unforeseen technical difficulties arose and some of the sections were not completed until recently.

Numerous star catalogs have been prepared during the present century. They serve a variety of purposes and interests. Some give, along with other data, the approximate positions, magnitudes, parallaxes, and proper motions of the bright stars. Others, like the *Astrographic Catalogue*, are of a general nature, while several are "fundamental," that is, they contain positions of the highest possible accuracy for a limited number of stars. There are also catalogs and lists of double stars, variable stars, and special features of stars such as parallax, spectral type, proper motion, and radial velocity. A large number of stars are therefore known by their positions in some of these catalogs, but many (*e.g.* Lalande 21185, Groombridge 1618, and Lacaille 9352) still retain the names of earlier catalogs.

CHAPTER III

THE SUN—PORTRAIT OF A STAR

Much of what we know and believe to be true about the stars is based on our knowledge of the sun. Every star, like the sun, is at once a ball-like mass of intensely hot gases, an immense thermonuclear reactor, and a broadcaster of light, heat, and other forms of energy. And every star, the sun included, has both specific and general traits. Though no star is precisely like another star, it can have certain characteristics in common with other stars. Once these individual characteristics are known, a star can be assigned to a particular group, type, or class. The majority of stars, in fact, can be classified into a few basic types.

The Photosphere

However, at first sight the sun appears to be unique. It easily surpasses all the other stars in brightness, it is the only star whose shape and surface we can see, and it has a profound influence on the earth and its environment. But the sun's qualities are special only because of its nearness to us. Its average distance from earth, roughly 93 million miles, is a stone's throw compared with the distances of other stars.

When seen through very dark glass the sun shows a uniform bright disk, perfectly round, smooth, and solid looking. This part, known as the photosphere or light-sphere, has a diameter of 865,000 miles and could easily contain well over one million earths. Appearances, however, are deceptive. Although the photosphere looks solid, it is a thin layer of gases. It merely represents the deepest level to which our vision can penetrate. The sun is made of gases right through to its center. The gases are opaque to an observer outside and therefore set a definite limit to the extent to which he can peer into them. The effect of this opacity is shown on photographs of regions near the sun's edge or limb, where it produces a decided darkening.

The photosphere, with a temperature of about 5,800 degrees centigrade, itself appears slightly dark in contrast with bright patches known as faculae. These can be seen in white or integrated light near the sun's limb, but are better observed in the violet light of calcium vapor. Faculae are usually found near sunspots, but may appear when spots are altogether absent. Slow to develop and decay they are thought to rise like intensely bright and broad hills above the general level of the photosphere, although none has been seen projecting beyond the sun's limb.

Solar Granules

High-resolution photographs, taken under good atmospheric conditions, reveal that the photosphere has a granulated appearance. The best photographs were obtained in 1959 when Stratoscope I, an unmanned balloon carrying a 12-inch telescope, took photographs of the sun at a height of some 80,000 feet. They show that the granulation is produced by a fine mosaic of bright granules or cells. These have irregular shapes, are separated from each other by dark lines, and range in size from about 1,000 miles across down to about 150 miles. They are all short lived, forming and vanishing within a few minutes, and by so doing give the photosphere a bubbling or boiling appearance. They are thought to be the tops of columns of hot gaseous material which rise from the deep interior and transport the sun's heat on the last stages of its journey into space.

Sunspots

The most striking feature of the photosphere is the well-known dark sunspots, first seen through the telescope by Galileo in 1610. Their general east-to-west movement across the sun's

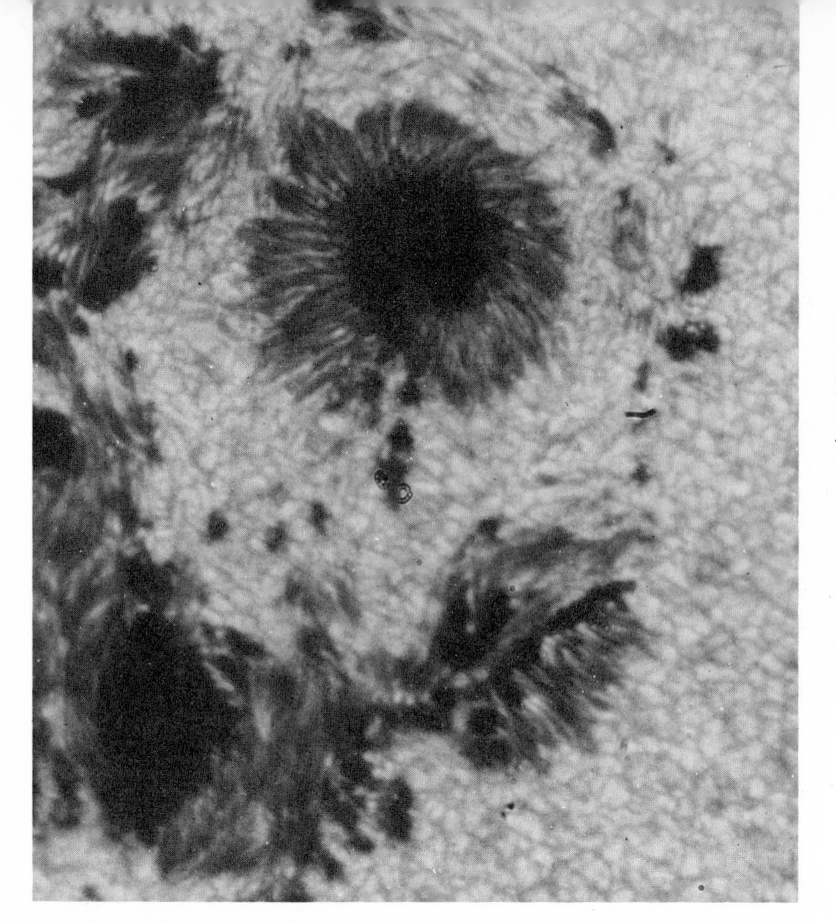

Sunspot group and solar granules photographed by the 12-inch telescope of Stratoscope I, 1959.
—PROJECT STRATOSCOPE OF PRINCETON UNIVERSITY, SPONSORED BY NSF, ONR, AND NASA

disk showed Galileo that the sun rotated once in about twenty-seven days. He claimed that since so large a body as the sun rotated, it was not unreasonable to propose, as Copernicus had done, that the much smaller earth had a similar motion. But his announcement that the sun had a blemished face was not well received by those who still favored the old idea that the "heavens" were perfect, eternal, and celestial. The discovery triggered off a bitter controversy, but as later results showed, it was really much ado about nothing. At the time everyone thought that because the spots looked dark they actually *were* dark. They are really quite bright, with temperatures between 3,000 and 4,000 degrees centigrade. They appear dark in contrast with the hotter, and therefore brighter, surrounding photosphere.

Sunspots have a great range in size. The smallest, called pores, are just a few hundred miles across and appear as dark specks. The largest occur as groups of spots and can extend over thousands of miles. At birth they look like dark spaces in the "sea" of granules. Most of these pores disappear, but those that grow in size tend to remain more or less round or "regular" in shape. Large spots, however, can acquire extremely complex shapes and structures, and are usually associated with smaller spots and pores. Groups of this kind occasionally grow to immense sizes and can survive for two and more solar rotations. The largest group ever observed occurred in 1947. In February it grew rapidly from a few small spots to a group visible to the unaided eye. By mid-March, in its second passage across the disk, it was much larger, and when it reappeared on March 30 for its third passage, it had an even larger area. At maximum development it covered 6,200 million square miles, or about 1 percent of the apparent area of the sun's disk.

Normally, two sunspots appear together in almost the same latitude. The westerly spot, acting as leader as the sun rotates, usually grows into a compact group dominated by a single spot. The easterly spot, trailing along behind, generally develops into a cluster of small spots which finally disintegrates while the leader is still prominent. The trailer tends to be farther from the sun's equator than the leader, a tendency which increases with the latitude.

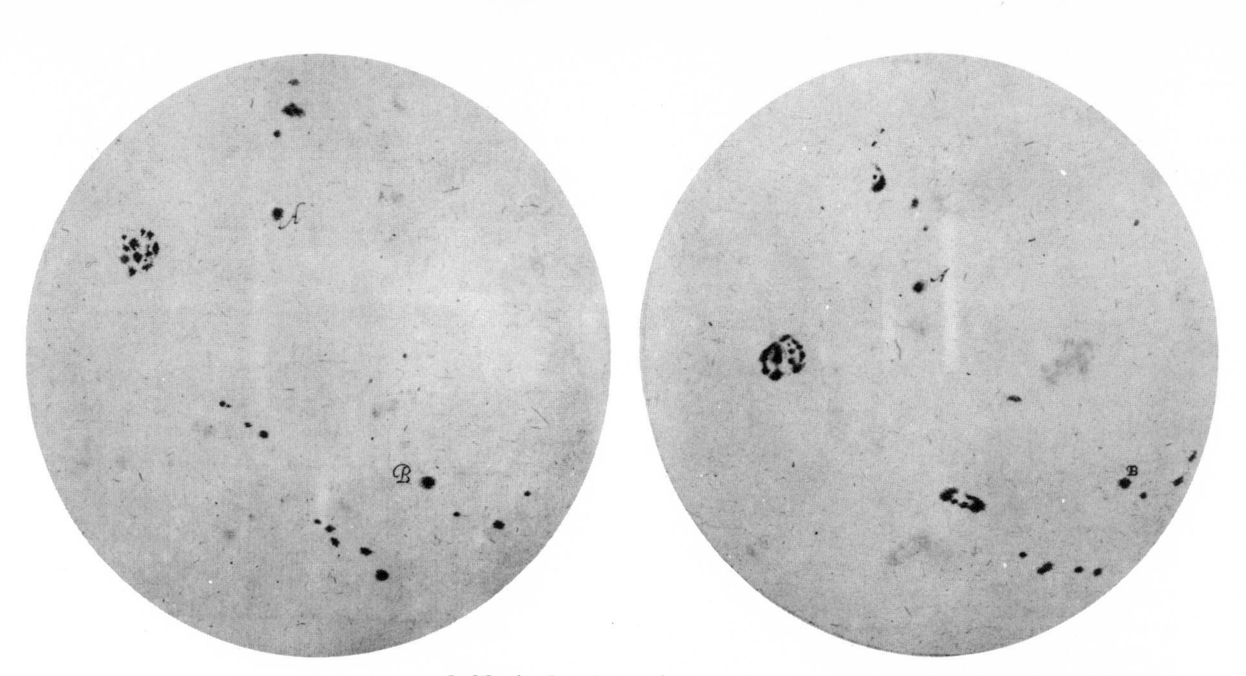

Galileo's drawings of sunspots on two successive days, from his work *Delle Macchie Solari.*

—Courtesy of the Trustees of the British Museum

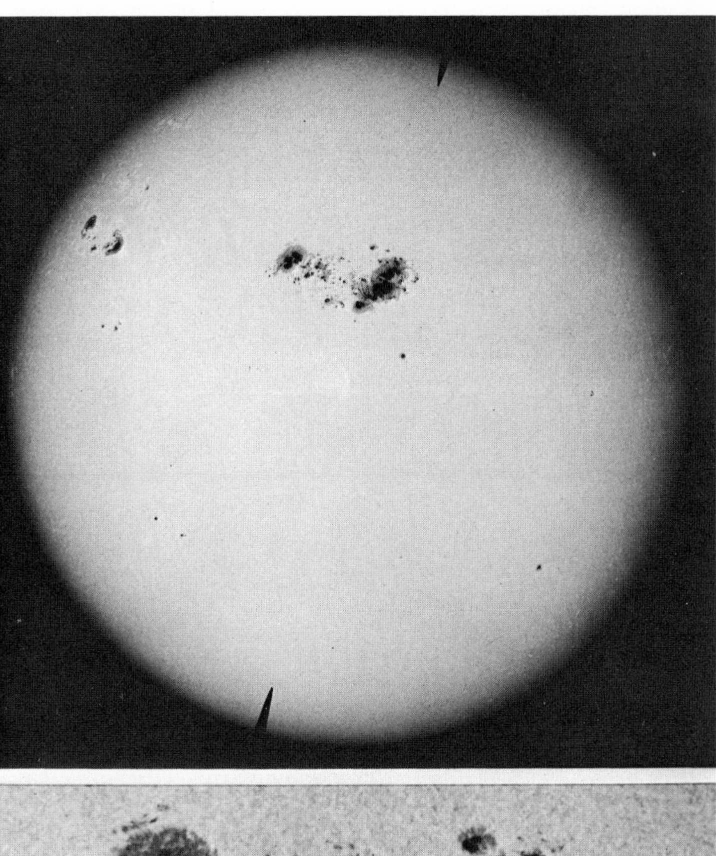

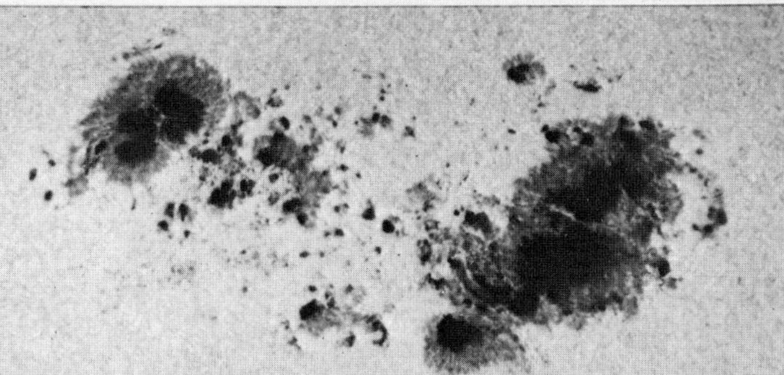

Great sunspot group of April 7, 1947.

—Mount Wilson and Palomar Observatories

A typical well-developed sunspot consists of two distinct parts, the central umbra and the surrounding penumbra. Usually the umbra is uniformly dark while the penumbra has a striated appearance and is less dark. In regular spots near the center of the disk the two parts are more or less circular and concentric. But as the spot approaches the limb its parts appear foreshortened and more and more oval in shape. Of course, if both occurred at the same level in the photosphere, one would expect the strip of penumbra nearer to the limb to become slightly narrower than that farther from the limb. This happens sometimes, but as a regular spot nears the limb, the side of the penumbra *nearest* to the limb is usually the last to disappear. The effect, named after A. Wilson, who first drew attention to it in 1774, indicates that these spots are funnel-shaped depressions. Some astronomers have even reported seeing spots on the limb as slight indentations, but their claims have not been generally accepted.

If sunspots were embedded in the surface of a solid sun their passage across the disk would enable us to determine one period of rotation for the sun. Measurements based on spots in all possible solar latitudes would all lead to just one period. Instead, they show that the rotation period at the equator differs from that at the poles and also that spots in any one solar latitude can change their relative positions. The sun cannot be a solid body. Its period of rotation is shortest at the equator (25.0 days) and increases with increasing latitude to reach 27.7 days in latitudes 45 degrees North and South.

These times, known as sidereal periods, are those of one complete rotation of the sun relative to the "fixed" stars. The sun, however, is viewed from the earth, which travels in the same direction as the sun rotates, and this leads to an apparent, or synodic, period of 26¾ days for spots on the equator. Sunspots are very rarely seen in latitudes higher than 45 degrees North or South, but spectrographic studies based on the Doppler effect show that the sidereal period continues to increase with increasing latitude and reaches about 34 days at the poles. Spot movements can also be used to determine

Sunspot near the sun's limb showing the Wilson effect.
—ROYAL GREENWICH OBSERVATORY

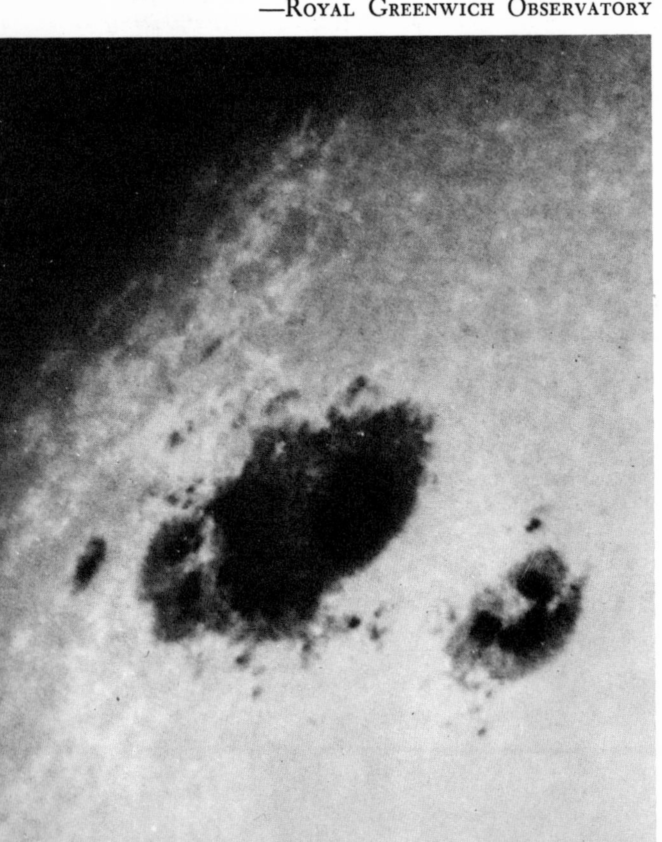

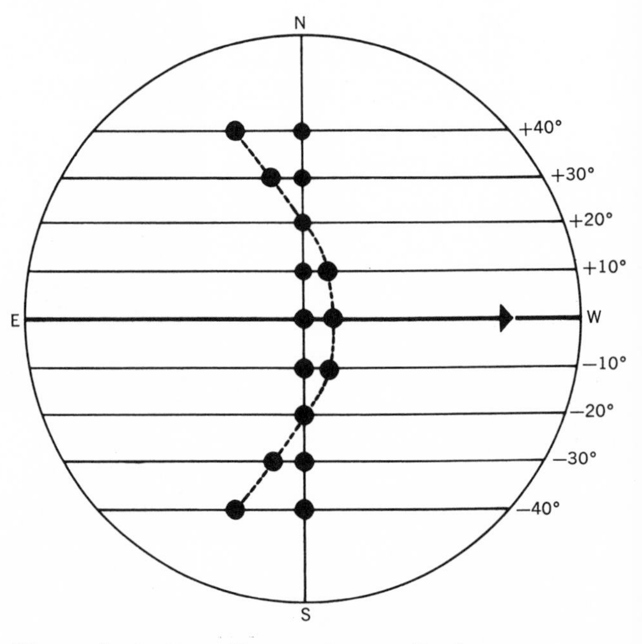

The sun's rotation. If sunspots were lined up on a single meridian on a given date, they would, after one complete rotation of the sun, lie on a curved line.

the direction of the sun's axis of rotation: this is such that the solar equator is tilted 7 degrees 15 minutes to the plane of the earth's orbit.

The most remarkable feature of sunspots is the way they rise and fall in number in a period known as the sunspot cycle. This cycle has an average value of 11.1 years but can be as short as 7 years and as long as 17. It is reckoned from a minimum, or the time when the spots are few in number or altogether absent, to the following minimum. At minimum the spots of a new cycle appear in belts about 35 degrees from the equator. They steadily increase in number to reach a maximum about 4½ years later. At the same time the latitude decreases, and at maximum the spots occupy belts about 15 degrees from the equator. The number then decreases, but so also does the latitude, and at the end of the cycle the spots reach an average solar latitude of about 8 degrees North and South. By this time a few spots of the next cycle have usually appeared in high latitudes, so the cycles often overlap. The date assigned to a minimum is very nearly that at which spots of the old and new cycles are equally numerous.

Although fairly continuous sunspot records

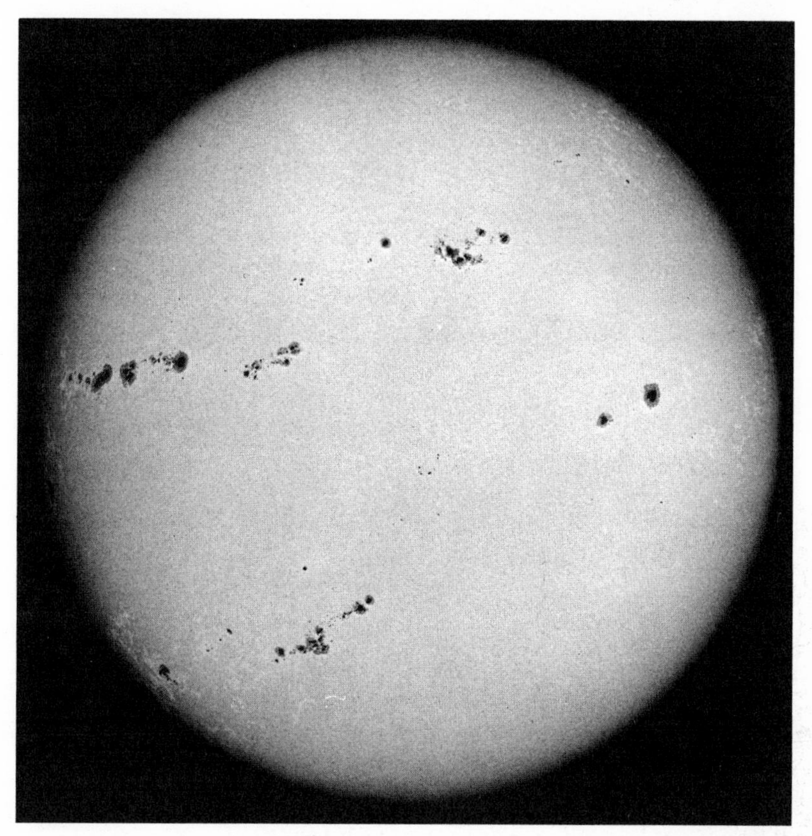

Photograph of sun at sunspot maximum December 21, 1957.
—MOUNT WILSON AND PALOMAR OBSERVATORIES

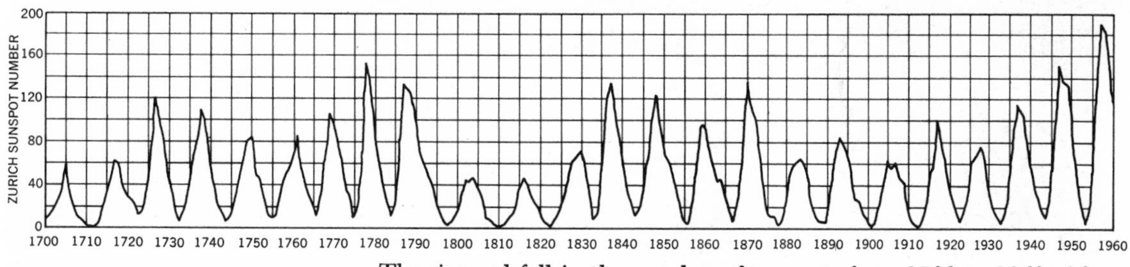

The rise and fall in the number of sunspots from 1700 to 1960. After M. Waldmeier.

have been kept since 1749, the task of evaluating and standardizing the early counts has not been easy. A study of all the available data shows that while the overall average length of a cycle is 11 years, the average length during the present century has been nearer 10 years. Short cycles tend to be associated with higher-than-average spot numbers—the 10-year cycle which began in 1944 yielded an all-time high of 3,420 groups.

In 1908 the American astronomer, G. E. Hale developed an instrument known as the spectroheliograph, which enabled him to photograph the entire visible disk of the sun in the

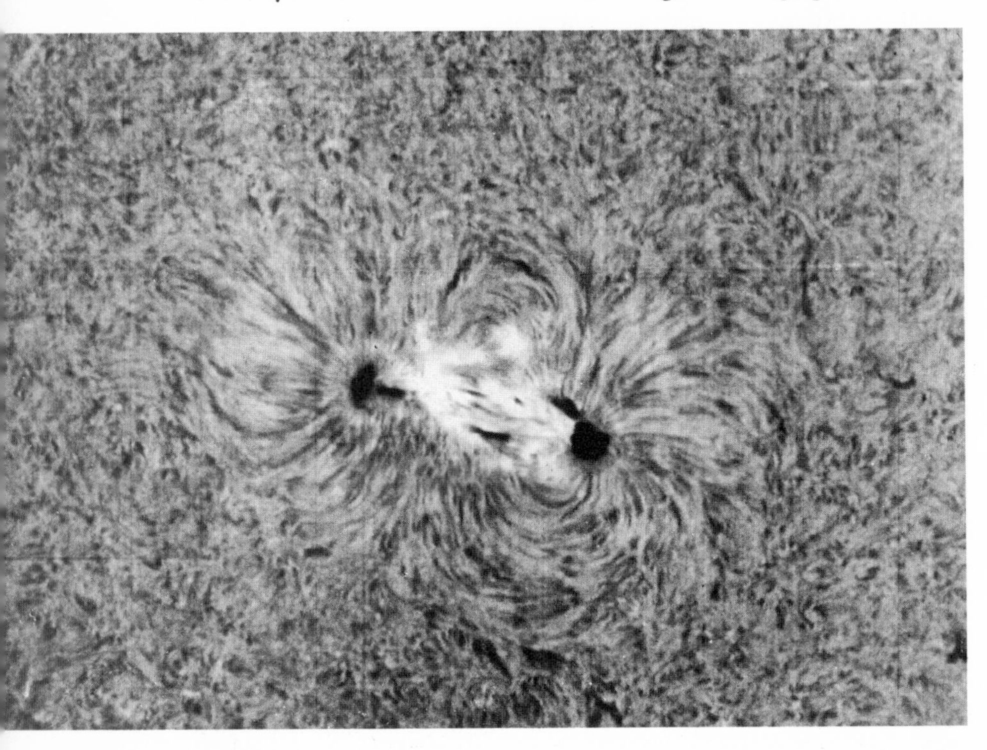

Bipolar spot group photographed in the light of hydrogen-alpha showing the whirlpool effect.
—Mount Wilson and Palomar Observatories

Funnel-type prominence photographed in the light of hydrogen-alpha.
—Sacramento Peak Observatory, Air Force Cambridge Research Laboratories

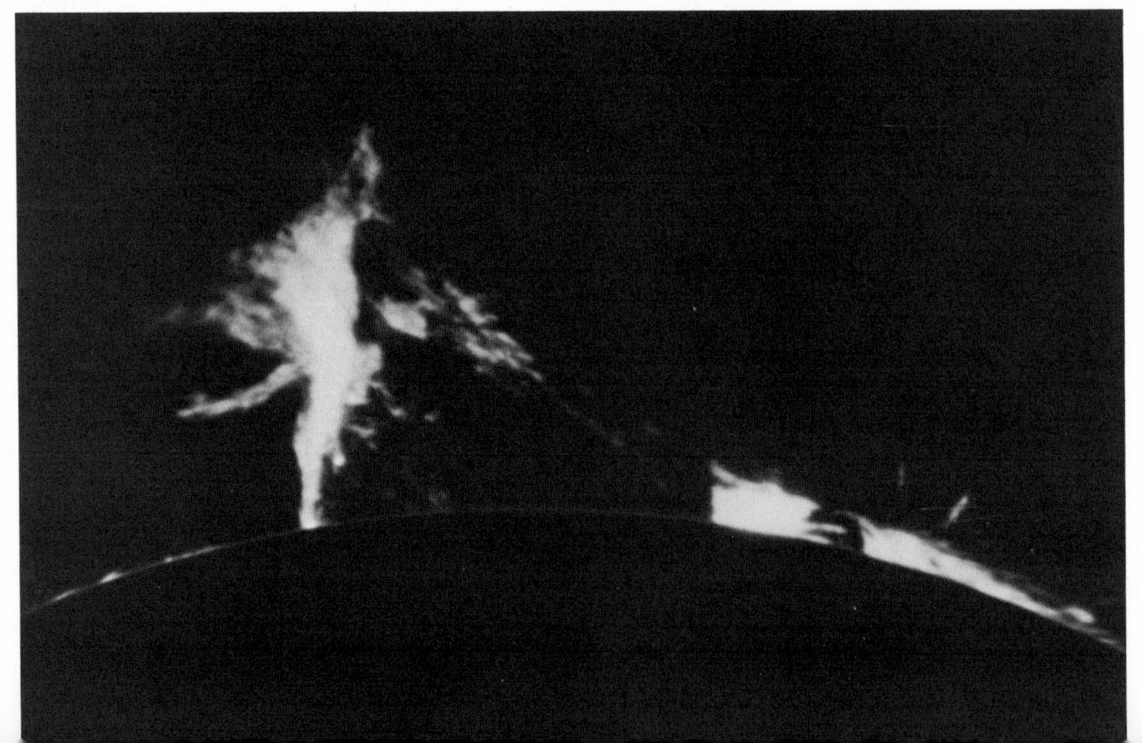

light of the red H-alpha line of hydrogen. The pictures showed that dark filaments located directly above the photosphere, formed a distinct whorl-like or whirlpool pattern around regular sunspots. Hale surmised that these spots had strong radial magnetic fields derived from the rapid revolution of electrically charged particles in the whirlpools. A ready method of testing this lay in the discovery by Zeeman that when a light source is placed in a strong magnetic field, the lines in its spectrum are split into two or more components. Peculiarities of this kind had already been noticed in the spectra of sunspots, and Hale had no great difficulty in proving that his theory was correct. His own and later investigations revealed that sunspot fields were invariably quite strong. They range from a few hundred gauss to about 3,800 gauss (in exceptional cases), whereas the strength of the earth's magnetic field is only about 0.6 gauss, which is very weak in comparison.

Hale found that the direction of a magnetic field, or its polarity, was opposite in the leading and trailing members of a sunspot group. This also held true for pairs with wide separations, indicating that all double-spot systems were bipolar. Even when only a single spot appeared, traces of the reverse polarity were sometimes found nearby. In 1908 the leading spots of groups in the sun's northern hemisphere had southern, or negative, polarity. But in 1912, when the first groups of the next cycle began to appear, the arrangement of their polarities was reversed. Leading spots in the northern hemisphere had positive polarity while their counterparts in the southern hemisphere had negative polarity. Similar reversals in polarity occurred at the minima of 1923, 1933, 1944, 1954, and 1965, thereby establishing beyond all

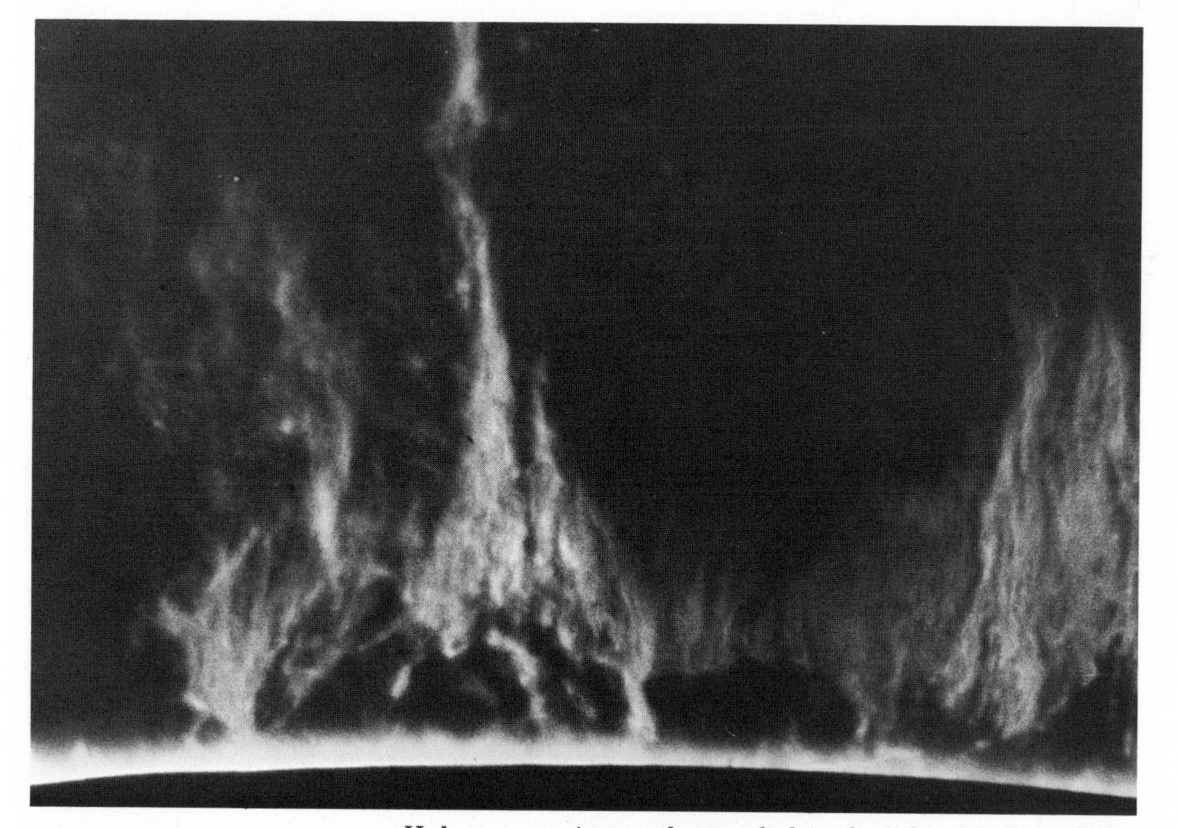

Hedgerow prominence photographed in the light of hydrogen-alpha.
—SACRAMENTO PEAK OBSERVATORY,
AIR FORCE CAMBRIDGE RESEARCH LABORATORIES

possible doubt that the period of complete change in polarity, or magnetic cycle, is twice as long as the average sunspot cycle.

The distribution of magnetic-field intensity in sunspot regions can now be studied in great detail by means of the magnetograph, an instrument that depends for its success on extremely high resolutions of the solar spectrum and ultrasensitive photoelectric detection devices. Its development is due largely to the pioneer work of H. D. and H. W. Babcock of the Mount Wilson and Palomar Observatories. Its records, called magnetograms, can show the distribution of magnetic field strength and polarity over the sun's entire visible disk.

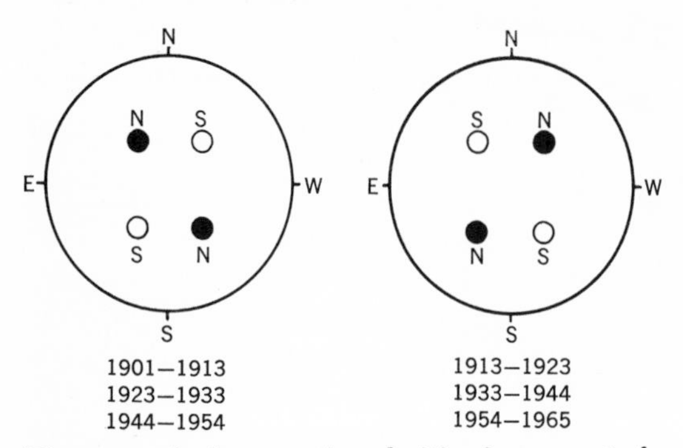

The reversal of magnetic polarities in sunspots for the last six sunspot cycles.

Solar Magnetism

Magnetograms made in 1955 revealed for the first time a widespread but weak solar magnetism, later to be known as the poloidal field. In 1955 the overall, or general, polarity of this field was positive in high northern latitudes and negative in far southern latitudes. The intensity, only one or two gauss, decreased during 1955–1956 and eventually could not be detected. Then, toward the end of 1957, the field in the southern polar regions reappeared with reversed polarity, and for nearly a year both polar regions had weak positive fields. Finally, in November 1958, the northern field reversed, thereby bringing about a situation precisely opposite to that found in 1955. These and subsequent observations show that the reversal of the poloidal field takes place about three years after a sunspot minimum. According to recent work by A. B. Severny of the Crimean Astrophysical Observatory, the field is not uniform but is composed of numerous magnetic regions containing small elements of different polarities. The sun qualifies as a magnetically variable star, but it is by no means unique in this respect. H. W. Babcock has found that some extremely hot stars have strong magnetic fields.

Theories of the Origin and Nature of Sunspots

Hale suggested that sunspots were funnel-shaped whirlpools formed by rapidly expanding, and therefore cooling, gases. In some cases the whirlpools associated with the spots of a bipolar group rotated in the opposite sense or direction as if to suggest that they marked the open ends of a great horseshoe electromagnet embedded in the sun. Later studies, however, showed that this was not always the case, nor did the direction of a whirlpool necessarily agree with the polarity of the magnetic field associated with it. Magnetic fields could not, therefore, be ascribed to whirlpool motions alone. Recent evidence indicates that although the fields are probably due to electric currents in the form of circulating electrons, their high intensities tend to slow down or inhibit the convection of hot material. Sunspots are not really whirlpools but islands of relative calm in an otherwise stormy sea.

None of the several theories of the origin and nature of sunspots is satisfactory, for none accounts for the 22-year magnetic cycle or explains why the spots rise and fall in number and at the same time drift toward the equator. One theory, originating with Bjerknes and modified by Walen, interprets the spots as regions where great toroidal, or doughnut-shaped, magnetic fields break through the photosphere. These fields, normally submerged in the photosphere, are supposed to writhe about like snakes, producing visible effects when they break through the surface. When they do break through, and when the field strength is high, the convective flow of material is inhibited and a relatively cool area develops to form a sun-

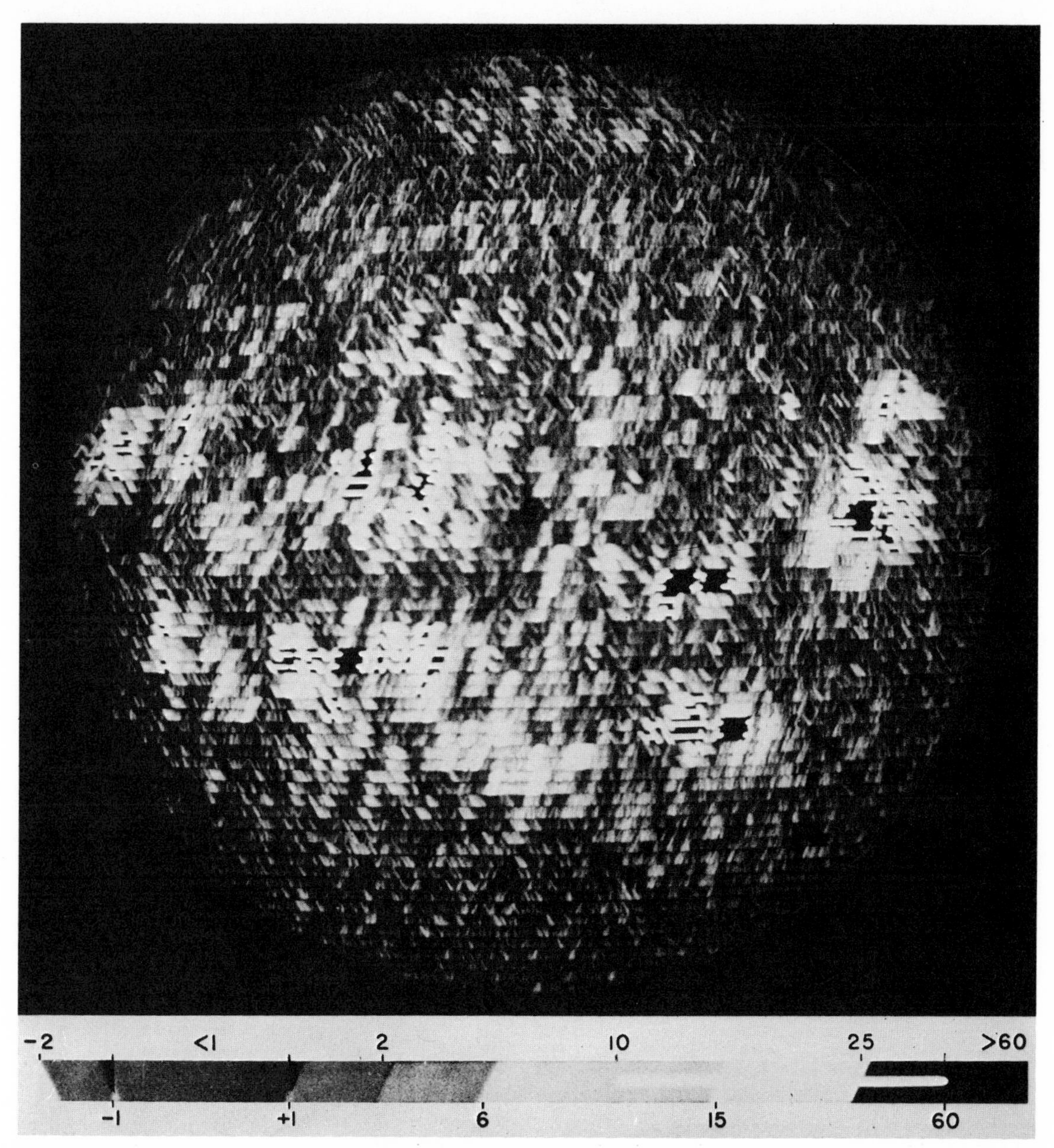

A magnetogram or magnetic map of the sun's disk showing the location, field intensity, and polarity of weak magnetic fields. The calibration strip at the bottom of the picture shows how the recording line slants to right or left to indicate magnetic polarity, and how it changes brightness and form to indicate seven different levels of magnetic field intensity in gauss. The extended magnetic areas on the solar disk are characteristically bipolar, and usually produce sunspots as well as other solar activity.

—Mount Wilson and Palomar Observatories

spot. Invariably the lines of force continue to keep in or near the toroid, so when this reenters the photosphere it forms another relatively cool area and hence the second spot of a bipolar pair.

The Chromosphere

Above the photosphere and the sunspots lies the chromosphere, a complex region of tenuous gases about 10,000 miles deep. Normally it

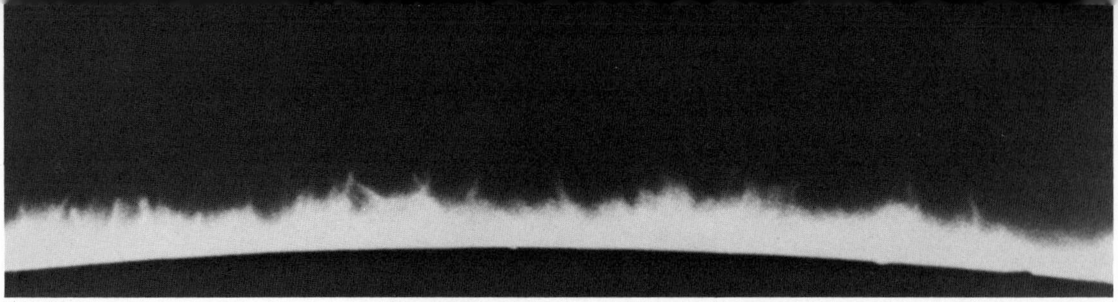

Spicules in the middle and upper chromosphere photographed in hydrogen-alpha light (with the 16-inch coronagraph at Sacramento Peak Observatory, Sunspot, New Mexico).
—SACRAMENTO PEAK OBSERVATORY,
AIR FORCE CAMBRIDGE RESEARCH LABORATORIES

cannot be observed at all, owing to the brilliant glare from the photosphere, but something of its nature is revealed when the sun is totally eclipsed by the moon. It then shines with a reddish hue because of the preponderance of glowing hydrogen gas. Since the chromosphere's radiations are monochromatic (that is, they have discrete wavelengths), it is best observed with instruments that can isolate the H-alpha line and the H- and K-lines of calcium vapor. These instruments are the spectroheliograph, the coronagraph (a special telescope invented by B. Lyot in which the sun can be artificially eclipsed), and highly selective filters known as monochromators. The last, like the spectroheliograph, enable the entire visible disk to be photographed in the light of a single wavelength. The chromosphere also emits radio waves shorter than 10 centimeters in length and can be detected with radio telescopes.

Gases in the extreme lower levels of the chromosphere are cooler than the photosphere and produce, by selective absorption, many of the well-known dark Fraunhofer lines of the solar spectrum. These lines have enabled astronomers to identify a large number of elements on the sun, (all of which are familiar on earth) and to obtain information about the density and temperature of the sun's atmosphere. At greater heights the atoms of certain light elements are found, along with others which are ionized, or stripped of one or more electrons. The fact that hydrogen, helium, and ionized calcium can exist high above the photosphere puzzled astronomers for years, and several explanations, none satisfactory, were advanced.

Solar Phenomena

SPICULES It is now thought that material and energy from the photosphere are carried high into the chromosphere by means of geyser-like columns of gas known as spicules. Their average diameter is about 500 miles and the gases forming them shoot upward at speeds of some 20 miles a second. They are probably connected with the granules and hence with convection currents below the photosphere, but proof of this association is lacking and the place and mode of their origin remains a mystery. Unfortunately the lower chromosphere, and therefore the lower parts of the spicules, lies too near the photosphere to be visible in a coronagraph. Those most easily observed stand out in profile above the sun's limb. Their moving spikelike tops then look like a field of wheat stirred by the wind or, as one early observer described them, the flames of a burning prairie.

PLAGES The sun's disk photographed in the light of the K-line has a curious mottled appearance. Bright, irregular patches, referred to as calcium flocculi or plages, occur around both faculae and sunspots. They are highly luminous clouds of ionized calcium vapor which soar into the chromosphere. Like the faculae, they indicate the great extent of the hot turbulent regions that surround the relatively cool and quiet spot areas. Similar photographs in the H-alpha line show a finer mottling. Bright patches, due to extremely hot hydrogen gas, are also invariably found in spot regions. In addition, dark threads are arranged around some spots, producing the whirlpool appearance discovered by Hale.

DARK FLOCCULI H-alpha photographs also show long, dark patches called dark flocculi. These are really hydrogen prominences, or great filamentous clouds of shining gas in the middle and upper chromosphere. They appear dark only when photographed in projection against the more brilliant solar background. When they happen to be on or near the limb their profiles

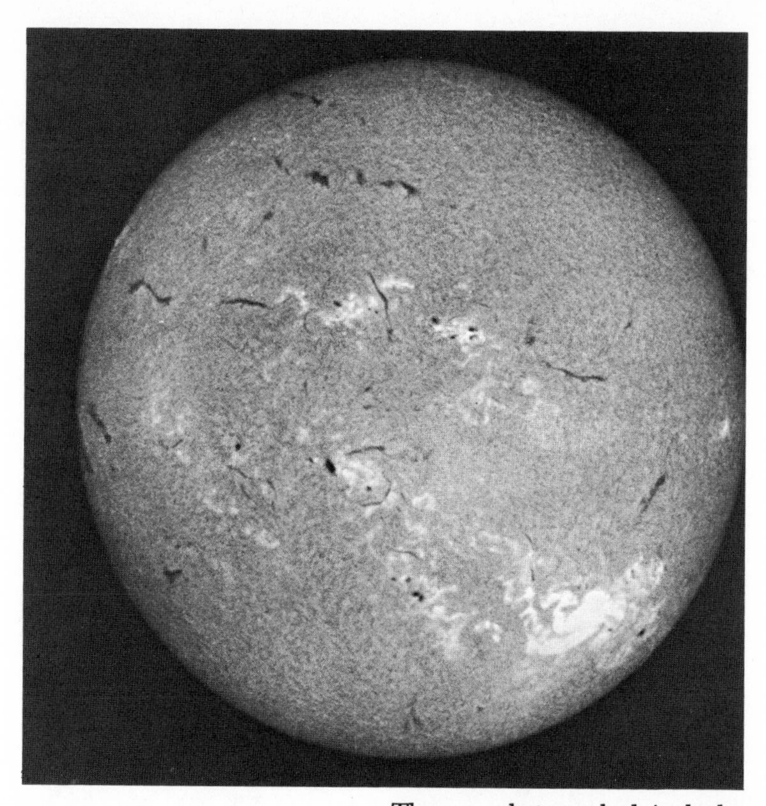

The sun photographed in hydrogen-alpha light showing bright and dark hydrogen flocculi and a large flare (bottom right). Sixteen-inch coronagraph at Sacramento Peak Observatory, Sunspot, New Mexico.
—Sacramento Peak Observatory,
Air Force Cambridge Research Laboratories

Surge-type prominence photographed in the light of hydrogen-alpha.
—Sacramento Peak Observatory,
Air Force Cambridge Research Laboratories

stand out bright against the dark sky. They then rise like gigantic tongues of fire high above the swaying field of spicules.

PROMINENCES Although prominences are best seen in the red light of hydrogen they can also be observed in the light of ionized helium, calcium, and other metals. Early studies suggested that there were two broad groups—quiescent and eruptive. The former changed slowly and sometimes persisted for many days. They often closely resembled trees, hedgerows, and arches, were rarely, if ever, associated with sunspots, and could occur at any time in any solar latitude. The eruptive variety developed and disappeared in a matter of minutes, often forming outward moving jets and explosive surges that underwent violent changes in both form and structure. They were usually associated with active sunspot areas and never appeared in very high latitudes.

Recent studies have led to a more extensive classification. Time-lapse motion-picture techniques enable us to follow in detail the complex forms and motions of prominences. The films show that there is certainly no clear-cut division between the two classes. Even the most quiescent prominences have internal movements. They can also suddenly surge upward and dissolve, as if blown apart by a great shock wave. A prominence can first become luminous at a great height and shed material sunward in graceful curves. Or it can take the form of an almost-complete loop in which gas streams both upward and downward as if controlled, or at least greatly influenced, by large-scale magnetic fields. Sometimes their parts attain velocities so great that they escape completely from the sun.

FLARES More violent than even the most active prominences are sudden explosive disturbances known as solar flares. They have their origin in plages which either surround complex sunspot groups or occupy regions where the magnetic polarities are both strong and opposite. In fact, they are thought to obtain their immense energies from the rapid collapse of magnetic fields, but just how the collapse takes place is unknown. Only the most outstanding flares are large enough to be seen in white, or integrated, light, and even these appear merely as small but intensely bright patches on the photosphere. Smaller flares are best observed in

Small loop prominences photographed on a large scale in the light of hydrogen-alpha.
—SACRAMENTO PEAK OBSERVATORY, AIR FORCE CAMBRIDGE RESEARCH LABORATORIES

Large prominence on the sun's limb during the total solar eclipse of May 29, 1919. Orientation: North-South axis runs upper right to lower left.
—ROYAL ASTRONOMICAL SOCIETY

H-alpha light. They are quite numerous and are comparable in intensity to larger ones. Most numerous of all are very small outbursts called microflares; as many as seventeen have been recorded in one hour.

The importance of a large flare lies mainly in the fact that it blasts out liberal doses of both wave and particle radiations, and these produce remarkable terrestrial effects. The wave radiations, in the form of X rays, ultraviolet rays, visible light, and radio waves, reach the earth in 8.3 minutes. The X rays and ultraviolet then react with the ionosphere to bring about short-wave radio fadeouts, the enhancement of atmospherics at long wavelengths, and disturbances of the earth's magnetic field. The particle radiations consist of beams of electrified atomic particles, mostly protons (the nuclei of hydrogen atoms) that have been accelerated to high energies in the flare. They reach the earth in times which depend largely on their energies, that is, on their velocities, always assuming, of course, that the earth happens to be in the path of the beam. The slowest particles make the

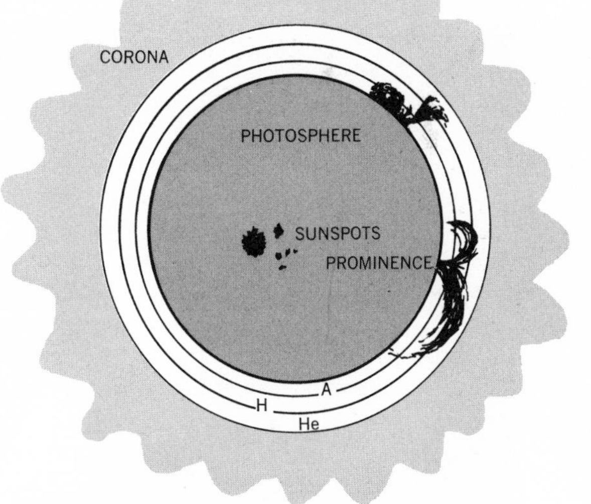

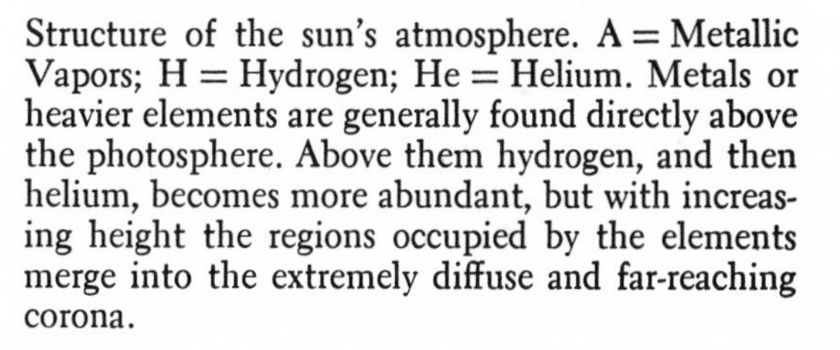
Structure of the sun's atmosphere. A = Metallic Vapors; H = Hydrogen; He = Helium. Metals or heavier elements are generally found directly above the photosphere. Above them hydrogen, and then helium, becomes more abundant, but with increasing height the regions occupied by the elements merge into the extremely diffuse and far-reaching corona.

The zodiacal light, photographed by D. E. Blackwell and M. F. Ingham at Chacaltaya, Bolivia, at an altitude of 17,000 feet, with the sun 18 degrees below the horizon. The bright streaks in the photograph are star trails.

—Royal Astronomical Society

crossing in twenty to forty hours, while the fastest take well under an hour. They then produce so-called magnetic and ionospheric storms with associated vivid auroral displays and severe disturbances to radio reception. Occasionally, when a flare is particularly intense, the particles travel at speeds approaching the velocity of light. They can then pass unhindered into the earth's upper atmosphere and are referred to as cosmic rays. These rays, solar in origin and strongly directional, must be differentiated from others that arrive from all directions in space and are clearly nonsolar.

The Corona

During a total eclipse of the sun the dark edge of the moon is bordered by a broad fringe or halo of pearly white light. This is the corona, or that part of the sun's atmosphere outside the chromosphere. But although it looks very striking and strangely beautiful, its brightness falls off rapidly with distance from the sun. Ground-based impressions, obtained during total eclipses, are restricted. Photographs taken from high-altitude balloons show that it extends to at least 13.5 degrees from the sun, a distance of about 12 million miles, while other studies indicate that it merges into a feature known as the zodiacal light. The latter is produced by sunlight scattered by dust particles distributed in the general plane of the solar system. Modern studies reveal that the corona reacts with the earth's atmosphere: it envelops the earth and probably fills the whole solar system.

The brightness of the corona is due largely

to sunlight scattered by a very tenuous and highly ionized gas. The scattering agents are electrons and protons that move freely in vacuumlike conditions and their great speeds give the gas an extremely high temperature. Close to the sun the average temperature of the corona is about 1 million degrees centigrade, and this, combined with the exceptionally low density, means that its atoms (mainly iron, nickel, and calcium) exist in conditions of high excitation. As a result the atoms are considerably ionized. An atom of iron, for example, normally has 26 electrons, but in the corona it could have as many as 13 electrons stripped off. These ionized atoms can emit a wide range of X rays, together with characteristic wavelengths in the ultraviolet and visible parts of the spectrum. They are responsible for bright emission lines in the coronal spectrum and so contribute, although to quite a small extent, to the light of the corona itself.

The corona is also a continuous emitter of radio waves. The waves are thermal in origin; that is, they are similar in origin to infrared, or heat, waves. They occupy a wavelength band of 1 centimeter to 15 meters and their steady everyday emission and reception is a normal feature of the so-called quiet sun. Because the wavelength decreases with increasing distance from the photosphere, radio astronomers can study the radio corona layer by layer merely by tuning their receivers to the appropriate fre-

The solar corona near the time of sunspot minimum. Photographed during the total solar eclipse of September 21, 1922.
—LICK OBSERVATORY

Loops in the sun's corona photographed in the light of the 5303 line of ionized iron on November 22, 1956. The sequence of nine frames covers a time interval of only 43 minutes.
—SACRAMENTO PEAK OBSERVATORY,
AIR FORCE CAMBRIDGE RESEARCH LABORATORIES

quencies. Recent observations of this kind, together with others of an optical nature, have revealed the great extent of the corona and also confirmed the conclusion that its temperature is extremely high.

At some wavelengths the radio emission changes in intensity. The changes are closely related to the occurrence of sunspot areas, but they nevertheless remain comparatively minor events. Much more striking are bursts in which the intensity rises above that of the quiet sun by hundreds and even millions of times. These come from localized regions in the corona and are associated with active sunspot groups, large solar flares, and occasionally with no visual phenomena at all. The radio emission associated with flares is probably synchrotron radiation, so named because it was first detected at optical wavelengths from a synchrotron in the laboratory. This is produced over a large wavelength range by electrons and other electrified particles when they are accelerated to extremely high speeds, and spiral along lines of magnetic force. Many cosmic radio sources are also emitters of this kind of radiation.

Near the sun the structure of the corona is decidedly complex. Photographs taken during total eclipses or with coronagraphs show numerous arched filaments and streamers. Their extent and prominence varies markedly with the number of sunspots. At sunspot minimum, long curved streamers from the sun's equatorial regions reach to great distances, while relatively short tufts appear at the poles. At sunspot maximum, on the other hand, the corona as a whole is much more compact and more or less uniformly distributed around the sun's limb.

How the corona can be so hot or, for that matter, exist at all, is one of many outstanding solar mysteries. There is undoubtedly a direct transference of energy and material from the chromosphere to the corona. One suggestion is that the carrying agents are acoustic shock waves, another, that they are magnetohydrodynamic disturbances, or shock waves involved with variable magnetic and electric fields. Granules and spicules are probably important links in the supply chain, while the turbulence they represent is associated with convective currents which flow outward from the sun's interior. These currents are thought to be confined to a region called the convective envelope, where the flow of radiation (in the form of gamma and X rays) from the deep interior or core is converted into heat and then carried to the surface.

How the Sun Shines

Because the material of the sun is held together by its own gravitational attraction, each layer can support the tremendous weight of the overlying layers only by exerting an enormous outward pressure. This pressure is of two kinds: pressure similar to that exerted by air in the tire of an automobile (gas pressure) and pressure due to the intensity of the outward flow of electromagnetic radiation (radiation pressure). If it were not for these supporting pressures the sun would collapse under the mutual gravitational attractions of its parts and become a much smaller and extremely compact body. The fact that it is so large, hot, and massive means that the supporting pressure and therefore the temperature of its core is extremely high.

We also know that the sun, in common with the majority of stars, consists largely of hydrogen. In its outer parts helium accounts for about 26 percent of the mass and all other elements for only 2 percent. This discovery, made about thirty-five years ago, suggests that the source of stellar energy is derived from the conversion of hydrogen into helium. Indeed, this synthesis, called a thermonuclear reaction, is the only known process which can account for the prodigious amounts of energy emitted by stars and for the steady rates of this emission over millions of years. Further, the extremely high temperatures and densities of stellar interiors favor hydrogen as a thermonuclear fuel.

THE PROTON-PROTON REACTION AND CARBON CYCLE There are two main thermonuclear processes in stellar interiors—the proton-proton reaction and the carbon cycle. In the former, four protons are eventually converted into the nucleus of a helium atom, while in the latter a similar result is produced after a proton combines with a carbon nucleus. The carbon, incidentally, is not used up but acts only as a catalyst.

Each process proceeds in stages, and in each of these mass is converted into kinetic energy and gamma rays. The conversion takes place in accordance with the relationship $E = mc^2$ formulated by Einstein in his theory of relativity. While the mass converted (m) is very small in amount, the square of the velocity of light (c^2) is a very big number. Hence the energy produced (E) has a large value. In the case of the sun the conversion proceeds at the comparatively slow rate (for a star) of 4 million tons a second. This corresponds to the conversion of about 570 million tons of hydrogen into about 566 million tons of helium.

Considerations based on theory show that the material near the center of the sun is subject to a pressure of about one billion tons per square inch and has a temperature of about 15 million degrees centigrade. Its density is about 110 times that of water, a not unrealistic figure in view of the sun's fairly high mean density (1.4) and the tenuous nature of its outer regions. Under these conditions the core material still behaves like a gas and its protons have sufficiently high speeds to produce thermonuclear reactions. The main process is the proton-proton reaction, with the carbon cycle probably playing a minor role. The latter becomes more prominent at higher temperatures and is therefore effective in stars hotter and more massive than the sun.

Age of the Sun

These considerations, and others to be discussed later, have an important bearing on the probable ages and future histories of the stars. The sun, it is thought, has been shining for about 7 to 8 billion years. Yet its store of hydrogen, about 30 percent by mass at the core, is sufficient to maintain its present rate of energy output for at least another 2 to 3 billion years. After that, important changes in size, temperature, and brightness will occur. The depletion of the hydrogen resources in the core is accompanied by the core's contraction, and this in turn results in a higher central temperature and density. Thermonuclear reactions in which helium is converted into carbon and other heavier elements may then occur. But when the nuclear reactions eventually cease, as indeed they must, the sun will have to rely on gravitational contraction alone for its supply of inner energy. This could keep it shining for some time as a white dwarf star but by then it will be well on the way to the graveyard.

CHAPTER IV

THE SUN'S NEAREST NEIGHBORS

A large number of stars have undergone cumulative changes in their relative positions due to proper motions. These changes are expressed in terms of angular distance per unit time (*e.g.*, seconds of arc per year), and when the corresponding distances are known, can be interpreted as transverse velocities, or velocities at right angles to the line of sight. For most stars the transverse velocity is only one component of their motion. Another component depends on their velocity toward or away from the sun. This second component produces a slight shift in the lines of their spectra relative

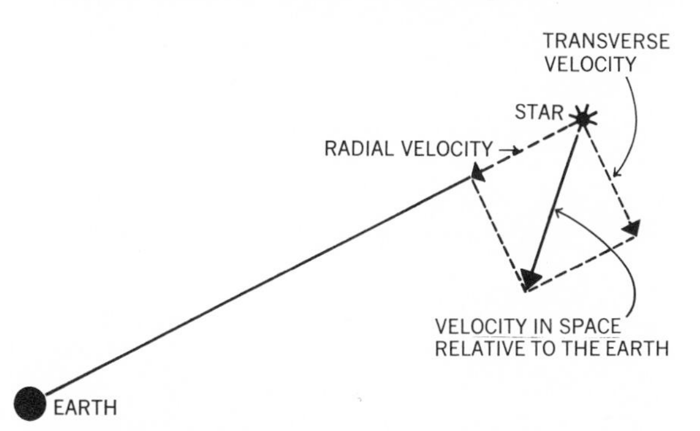

The two components of a star's velocity relative to the earth.

to the positions of the spectrum lines of a source at rest. The shift is known as the Doppler effect and its amount is a measure of a star's line-of-sight motion, or radial velocity. So, after allowance has been made for the orbital motion of the earth, the resultant of the two components is the actual velocity of a star relative to the sun.

In practice, the displacements due to proper motions are tangled up with those due to parallax, though their separation is not too difficult a task since proper motion is cumulative and parallactic motion is periodic. Both motions furnish clues to a star's distance, but the only safe criterion is trigonometric parallax.

The Nearest Stars

Thirty stars are now known to have parallaxes equal to or greater than 0.252 second of arc. They lie within a sphere of radius about 13 light years centered on the sun and are the sun's nearest neighbors. Twenty of them are below naked-eye visibility and only four are of the first magnitude. Most of the bright stars lie well outside this region. Capella, for example, a star of the first magnitude, has a parallax of only 0.073 second of arc and a distance of about 45 light-years. This is about the smallest angle of parallax that can be reliably measured by modern techniques. In fact, it would be more correct to say that angles of this order of size are estimated rather than measured. Parallaxes smaller than this are much more difficult to estimate, and below about 0.014 second of arc, or beyond a distance of 250 light-years, the uncertainty becomes so great that the technique ceases to be effective.

Once a star's distance is known, its apparent brightness can be used to determine its intrinsic brightness, or luminosity. The apparent brightness of a star or, for that matter, of any point source of light, decreases as the square of its distance increases. Thus if Capella were at twice its present distance it would be only a quarter as bright as it is now. At three times its present distance it would be only one ninth as bright, and so on.

Absolute Magnitude

An alternate way of describing the luminosities of the stars is to determine what their magnitudes would be if they were all at one standard distance from the sun. The distance chosen is 10 parsecs. Distances of 10, 100, 1000 parsecs therefore correspond to parallaxes of 0.1, 0.01,

0.001 second of arc respectively. The magnitude which a star would have at a distance of 10 parsecs is called its absolute magnitude. That of the sun is 4.9, of Sirius, 1.4, and of Wolf 359, 16.5.

In general the luminosity of a star depends on the temperature and area of its surface. We can demonstrate this by the homely example of two pokers of different sizes placed in a fire. At first they will turn dull red in color and then, if the heat is intense enough, eventually become white hot. They are much brighter in the latter condition than in the former. But although the brightness per unit area is the same for both pokers at a particular temperature, the larger one emits more light than the smaller solely because it has the larger surface area.

DETERMINATION OF SURFACE TEMPERATURES A similar situation arises with stars. As the temperature increases, the wavelength at which the radiation is most intense shifts progressively from the red end of the spectrum to the blue end. The temperature of a star's surface can therefore be deduced from its color. In practice the magnitude of a star is determined in the yellow and then in the blue by means of optical filters and a photomultiplier cell. The difference between the two magnitudes (blue minus yellow) is a "color index" which increases from negative values for hot blue stars to positive values for cool red stars. The temperature can

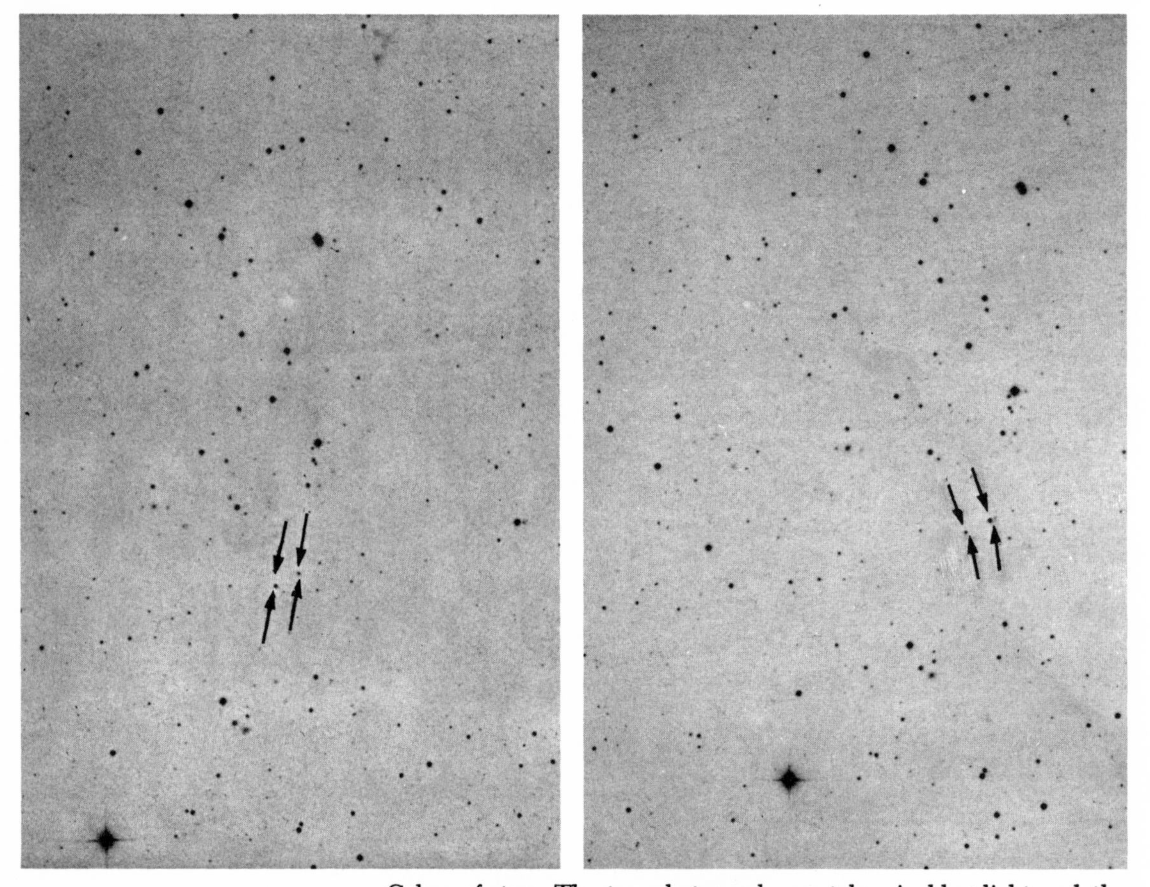

Colors of stars. The top photograph was taken in blue light and the bottom photograph in red light. The difference in the sizes of the images of the two stars marked shows that the left-hand star is blue and the right-hand star is red. Color indicates surface temperature.
—W. J. LUYTEN

then be derived from the color index by assuming that the star is a "black-body" or "perfect" radiator.

THE HARVARD SEQUENCE OF STELLAR SPECTRA Surface temperatures can also be determined by interpreting stellar spectra. The dark absorption lines give information about the physical states of the atmospheres, in particular of temperatures and densities rather than of chemical compositions. The study of stellar spectra shows that the majority of nearby stars can be classified in a continuous series or sequence of spectral types. These types are designated quite arbitrarily by the letters O, B, A, F, G, K, M, R, and N in this order form a sequence of changing color and also of progressively decreasing temperature. Each type has decimal subdivisions 0 to 9, so arranged that G0, for example, represents a slightly higher temperature than G2.

The sequence is known as the Harvard sequence, since it was derived from a spectrographic survey of 10,351 stars made by E. C. Pickering and his coworkers at Harvard College Observatory, Massachusetts. The spectra were cataloged in 1890 as the first *Henry Draper Catalogue*, which was followed by a much larger catalog of no less than 225,000 spectra.

General Characteristics of Different Stellar Spectral Types

Through spectral types O, B, and A the surface temperature decreases from over 30,000 to about 8,000 degrees centigrade and the color changes from blue to white. The maximum intensities of the radiation of O- and B-type stars lie in the ultraviolet. An important feature of their spectra is dark absorption lines due to helium and hydrogen. The former grow in intensity to reach their greatest prominence in type B2 and then weaken to disappear in type B9. The latter increase in strength up to type A2 and then gradually fade, almost to disappear

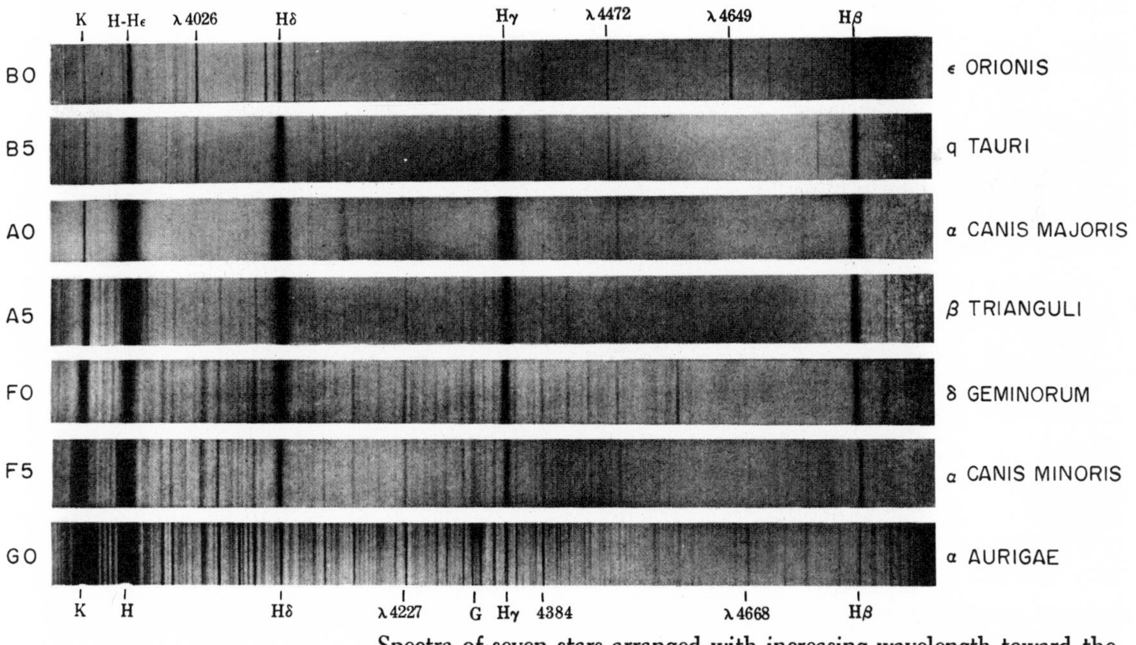

Spectra of seven stars arranged with increasing wavelength toward the right. The letters at the left denote spectral types and the numbers beside them the spectral sub-types. Differences in the relative intensities and number of the spectral lines show that the order B0 through to G0 is one of decreasing surface temperature (35,000 to 6,000 degrees centigrade).

—ROYAL ASTRONOMICAL SOCIETY

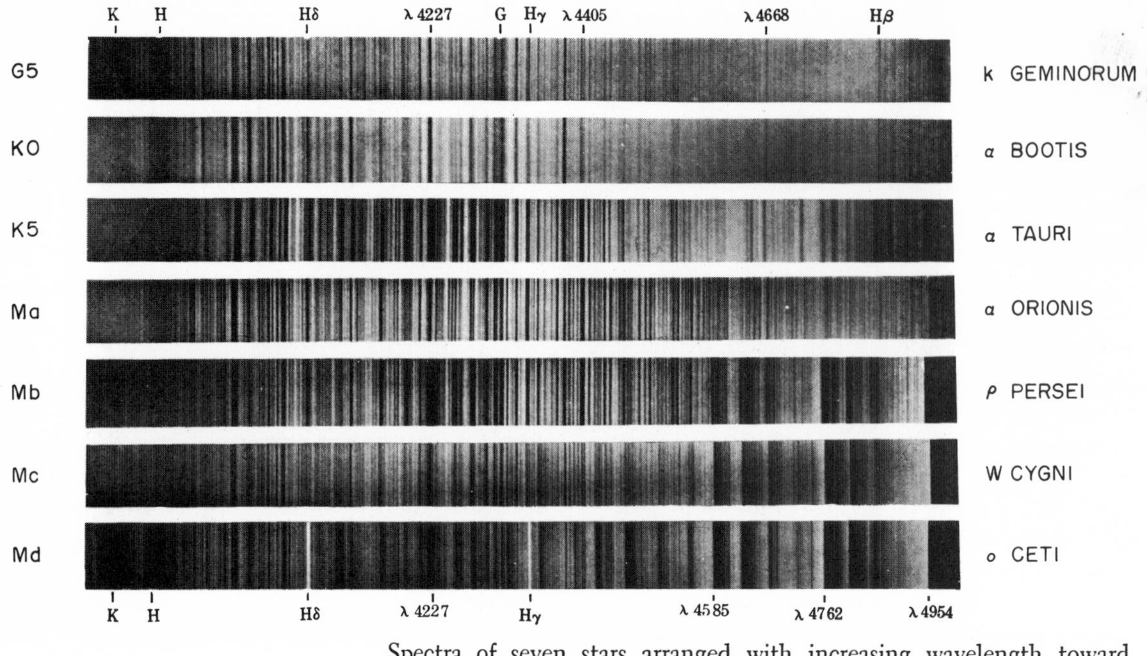

Spectra of seven stars arranged with increasing wavelength toward the right. The letters at the left denote spectral types and the numbers beside them the spectral subtypes: the suffix "e" indicates that the spectrum has emission lines in addition to absorption lines. In this series absorption lines due to metals are profuse and those due to molecules become increasingly prominent with decrease in temperature. The range in temperature from G5 through to M8 is from about 5,500 to about 3,400 degrees centigrade.

—Royal Astronomical Society

in type M. In type B8 some lines due to ionized metallic atoms such as iron, magnesium, and calcium begin to appear. These grow in intensity to become the chief feature of type A5.

The decrease in surface temperature continues through types F, G, and K. It falls from about 8,000 to about 3,500 degrees centigrade and the color changes from yellow-white to orange. Low-temperature lines due to neutral metallic atoms begin to appear and grow in number and strength through to type K. Molecular bands due to titanium oxide and hydrocarbons appear for the first time at type K5. The hydrogen lines continue to decrease in intensity but those due to ionized calcium become a prominent feature of type G. The sun is a typical example of type G2.

Through types M and N the surface temperature falls below 3,500 degrees centigrade and the color changes from orange-red to deep red. The maximum intensity of the radiation moves progressively into the infrared. In type M the low-temperature metallic lines and titanium-oxide bands are strong, while type N has bands due mainly to molecules of carbon monoxide (CO) and cyanogen (CN).

When the thirty nearest stars are arranged in order of luminosity an interesting fact emerges: only four—Sirius, Altair, Procyon, and Alpha Centauri—surpass the sun, and of these Sirius heads the list with a luminosity about 26 times that of the sun. All four are first-magnitude stars, but as we shall see presently, three of them consist of more than one star. Nineteen of the rest are considerably less luminous than the sun. They are no more than glowworms compared with the sun's electric-light brilliance. Most of them are well below naked-eye visibility and appear faint because they *are* faint. Faintest of all is Wolf 359, magnitude 13.5, with a luminosity about 0.000023 that of the sun. It would take some 430,000 stars each equal in luminosity to Wolf 359 to replace the sun in brightness.

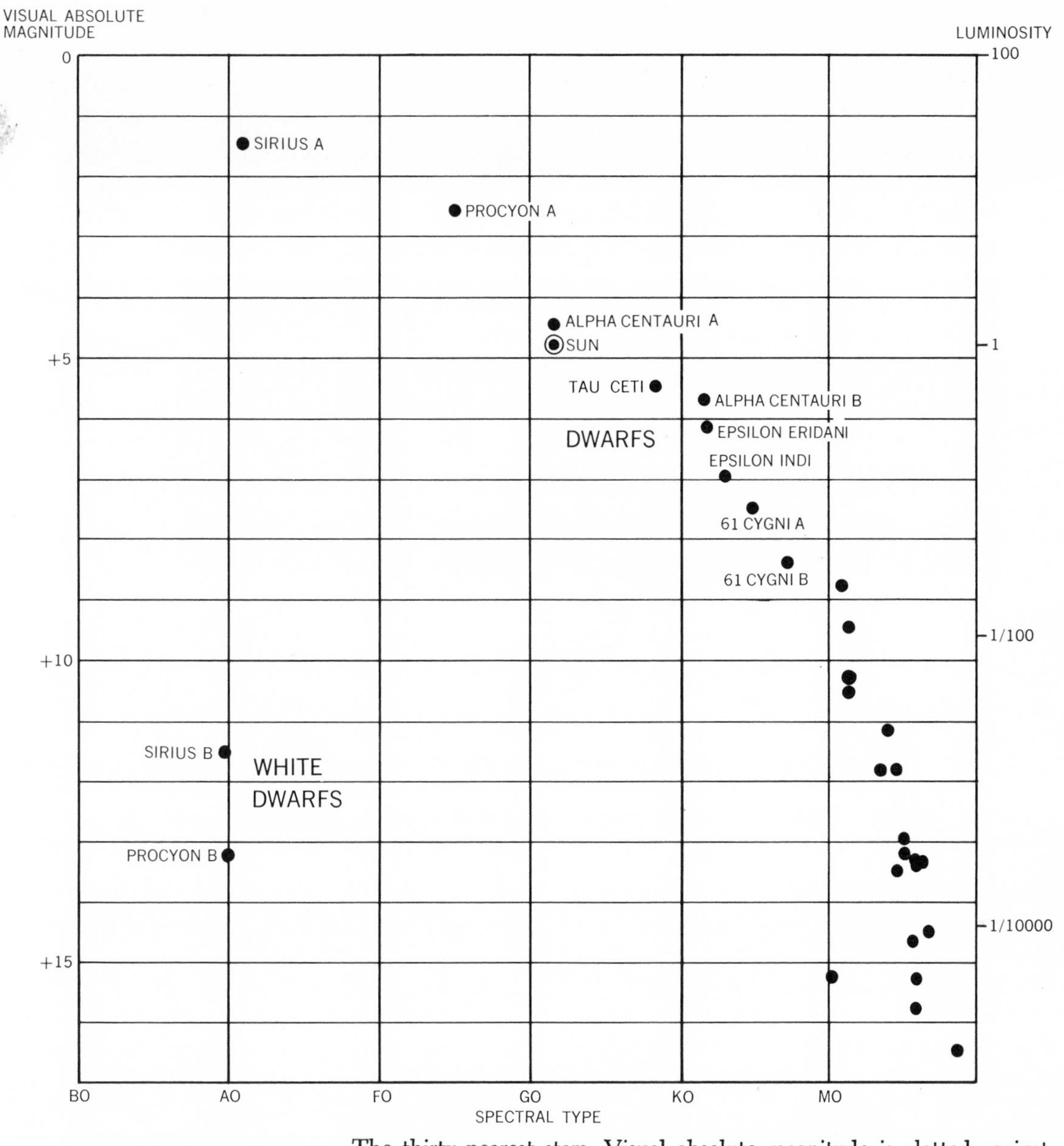

The thirty nearest stars. Visual absolute magnitude is plotted against spectral type.

The M-type spectra of these exceptionally faint stars indicate surface temperatures of about 3,000 degrees centigrade. They are therefore considerably cooler and redder than the sun, and in view of their very low luminosities, must have small surface areas. Theory provides some of them with diameters of only a few thousand miles. These particular stars, then, would thus be similar in size to the smaller planets. They are called red dwarfs, or sometimes M-type dwarfs.

Binary Stars

Another interesting feature of the thirty nearest stars is the fact that fourteen of them belong to pairs of stars known as binaries, while three are members of a triple system. In systems of this kind the component stars travel around their common mass-centers, or centers of gravity. They reveal their gravitational association by their periodic changes in relative position, although in some cases the changes are

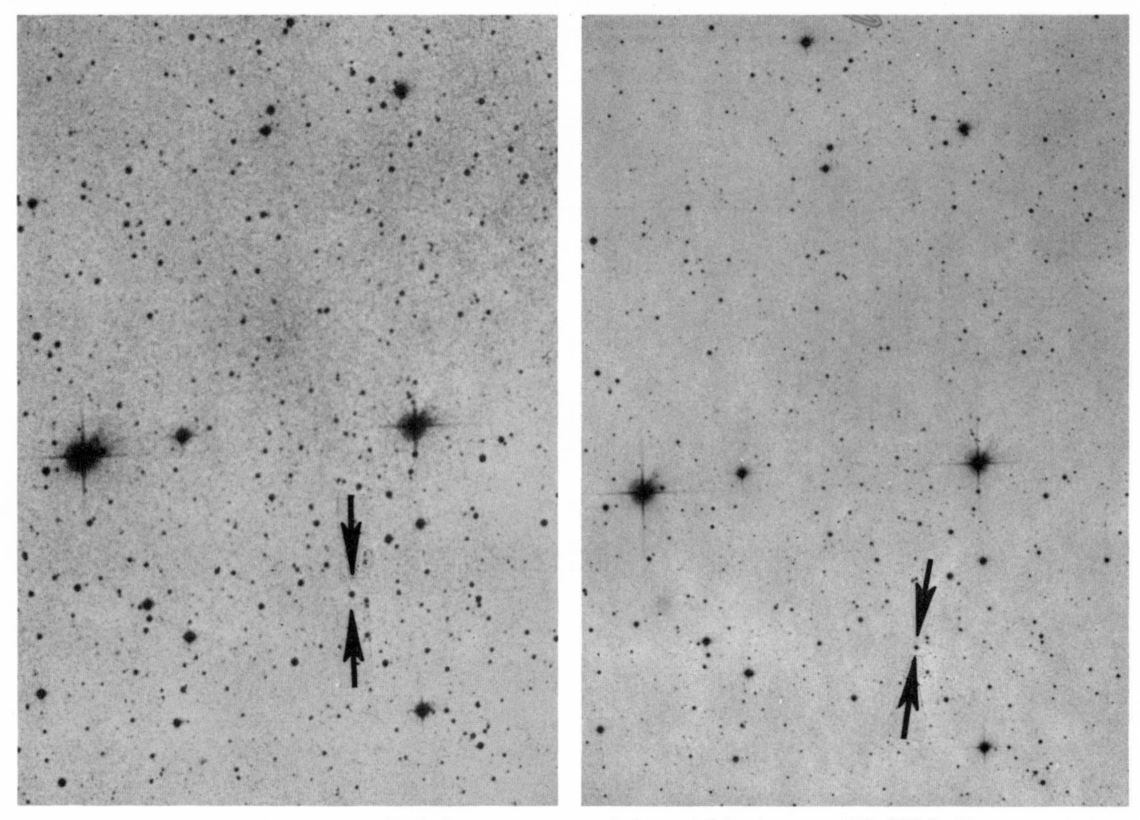

Relative movement of the neighboring star LP 658-2. Two comparison photographs, one taken in 1950 and one taken in 1963, show that the star has moved relative to the stars around it.

—W. J. Luyten

extremely slow. A binary star is similar in appearance to an optical double for both phenomena appear as one star to the naked eye. But the two are quite different in nature; the "components" of an optical double merely happen to lie in almost the same direction. The binaries *are* components of the same system.

61 Cygni One of the nearest binaries is 61 Cygni, a system of two K-type stars whose larger proper motion tempted Bessel to attack the problem of determining its parallax. To the unaided eye the components are well separated but faint, each being of about the sixth magnitude. In reality the brighter star, 61 Cygni A, is 19 times less luminous than the sun and is separated from its companion by about 80 A.U., or nearly 7,500 million miles. The system has been observed long enough to obtain information about the orbits of the two stars and to establish the fact that they revolve around each other in a period of about 720 years. It has thus been possible to investigate the system dynamically and determine the individual masses of the two stars. A is found to be slightly more massive than B, and the two together just barely outweigh the sun.

In 1943 the American astronomer K. A. Strand announced that 61 Cygni had an unseen companion. Small periodic variations in the motion of B indicated that a perturbing influence was at work in the form of a third body of small mass. Calculations then showed that this body was about 70 times less massive than the sun. This, coupled with the fact that the companion could not be detected on photographs taken with even large telescopes, indicated that it might be nonluminous and therefore planetary.

Similar variations have also been detected in the motions of Lalande 21185 and Barnard's star, two other near neighbors of the sun. Lalande 21185 is a red dwarf about 1/200 as bright as the sun and just over 8 light-years away. The unseen companion must be considerably less luminous than its primary, and this implies small size and mass. Barnard's star, another red dwarf, is about 6 light-years away. It was discovered in 1916 by E. E. Barnard, who noticed its exceptionally large proper motion of 10.3 seconds of arc per year. At present it is drifting northward in the constellation Ophiuchus and moves through an angle equal to the apparent diameter of the moon in about 180 years. The perturbing body is thought to have a mass only 0.0015 that of the sun, or 1½ times the mass of Jupiter.

Another unseen companion is associated with 70 Ophiuchi, a binary star about 165 light-years away. Its mass, about 12 times that of Jupiter, is about equal to that of the dark companion of 61 Cygni B.

If a wholly gaseous body had a mass similar to that of any one of these unseen companions it would be unable to maintain a central temperature high enough to bring about thermonuclear reactions. So if the masses are at all correct, the companions are not stars at all but planets which shine by reflected light only. This raises the intriguing thought that there may be many more bodies of a similar kind as yet undetected. Some stars may have more than one dark companion and perhaps even a retinue of lightweight planets.

Another nearby binary system is Alpha Centauri, 4.3 light-years away and the third brightest star in the night sky. To the unaided eye it looks like a single object, but a small telescope shows two stars, close together and well matched in brightness. The brighter component, of spectral type G2, is similar to the sun in luminosity, mass, and surface temperature. If the sun were placed alongside there would be no great difference between the two stars. The fainter component, of spectral type KI, is less luminous, less massive, and cooler than the sun.

About two degrees away is a third member, Proxima Centauri, a red dwarf flare star of apparent magnitude 10.7. Its parallax (0.762 second of arc) shows that it is slightly nearer to us than Alpha Centauri and therefore has the privilege of being the sun's next-door neighbor. It shares the same space motion as Alpha Centauri, but the shape of its orbit and period of revolution are as yet unknown.

FLARE STARS Another unusual binary star is L726–8, discovered in 1948 by W. J. Luyten of the University of Minnesota. Its distance is 8.6 light-years, so it qualifies as one of the sun's near neighbors. Each star, a small red dwarf of about the thirteenth magnitude, has a surprisingly low mass of roughly 1/25 that of the sun. Late in 1948 Luyten observed a short-lived increase of two magnitudes in the brightness of the fainter component, now known as UV Ceti. In other words, the star increased in luminosity sixfold. Two other red dwarfs, WX Ursae Majoris and Ross 882 had already been observed to behave in a similar way. A close watch was kept on all three stars and it soon became evident that the sudden increases in brightness were due to localized outbursts similar in nature to solar flares. In the case of UV Ceti they were far more violent than anything the sun has produced: on one occasion the star increased its luminosity about 250 times.

RADIO STARS By 1958 we knew that flares of one or two magnitudes occurred on UV Ceti at the rate of one every 35 hours. This led Sir Bernard Lovell of Manchester University to suspect that it might be possible to detect the associated radio emission with the 250-foot Jodrell Bank radio telescope. In the course of five years he and his associates succeeded in recording radio emissions not only from UV Ceti but also from four other flare stars. In the same interval all five stars were systematically photographed with cameras used in the satellite tracking network of the Smithsonian Astrophysical Observatory in Washington. Analysis of the data showed that the radio emissions occurred during the times of the flare outbursts. Obviously the sun is not the only star to be afflicted with flares, nor is it the only radio star. Of the nineteen red dwarfs in the sun's neighborhood at least three are flare stars—UV Ceti, Ross 154, and Krüger 60 B. Many more have been detected farther afield, but so far they have been found to be M-type dwarfs or, more rarely, K-type dwarfs.

SIRIUS In 1834 Bessel detected periodic variations in the proper motion of Sirius and suspected that it had an unseen companion. The variations, being periodic, enabled astronomers to predict the positions of the companion relative to its primary, but for many years attempts to detect it met with no success. In 1862, however, the American telescope maker A. Clark, while testing a large telescope on Sirius, saw a fairly faint star in almost the precise position predicted for that year. Actually the companion, Sirius B, with a magnitude of 8.5, is not particularly faint. It can be seen in telescopes of quite modest size when the eye is shielded from the brilliance of Sirius A.

The spectrum of Sirius A differs from the sun's in that the lines due to hydrogen and various ionized metals are more prominent. Their number and intensities indicate a surface temperature of about 11,200 degrees centigrade and are typical of spectral type A1. This higher temperature, considered in relation with the luminosity (26 times that of the sun), reveals a diameter which exceeds the sun's by some 80 percent.

SIRIUS B Sirius B, on the other hand, has a luminosity about 100 times smaller than that of the sun. Yet its color index shows that the surface temperature is about 10,000 degrees centigrade. It is white in color, but its A-type spectrum has a number of anomalies that show that conditions other than those of temperature and pressure are involved. The low luminosity and high temperature denote a diameter only

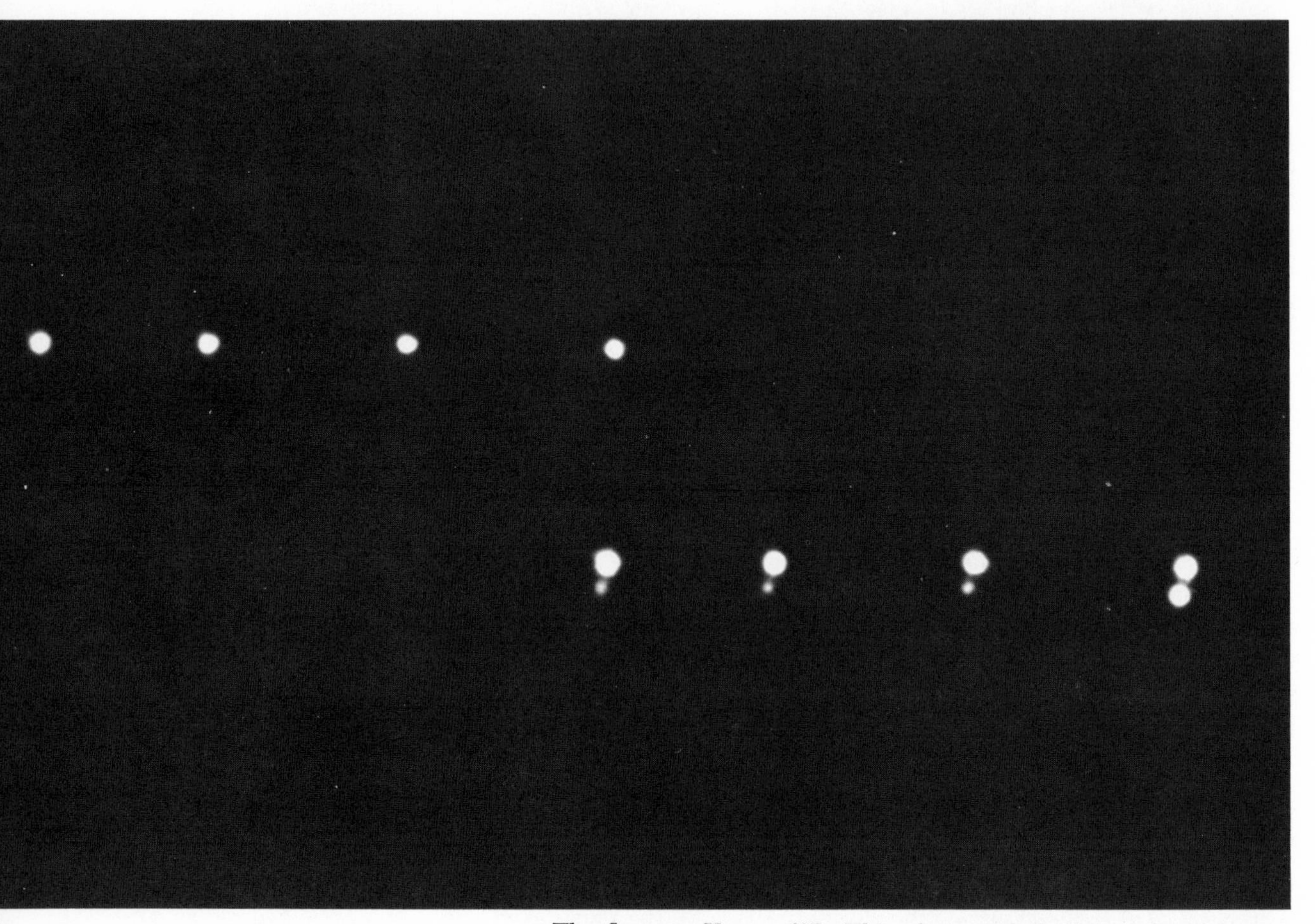

The flare-star Krüger 60B. This photograph is the result of five exposures of the binary Krüger 60 and a neighboring field star and shows the B component in the act of flaring (extreme right).
—SPROUL OBSERVATORY

1/50 that of the sun, or about 2½ times that of the earth. But the most remarkable feature is the mass, which can be determined from the size and shape of its orbit and the period of revolution (50 years), found to be about equal to that of the sun. A star so small in size yet so large in mass must have an extremely high average density. The estimated value is 500,000 times the density of water, so if a cubic inch of material could be scooped from the interior of Sirius B and transferred to the earth's surface it would probably weigh over 8 tons.

Further evidence for the high density and small size of Sirius B is provided by a small displacement of its spectral lines toward the red. A shift of this kind was predicted by Einstein, who found that it would vary directly with a star's mass and inversely with its radius. The shift was first detected by W. S. Adams in 1915 in the spectrum of Sirius B, and the observations, although difficult, were found to agree fairly well with theory.

Red-shift measurements for another star, 40 Eridani B, the first white dwarf to be discovered spectroscopically (in 1914), have recently given similar results.

White Dwarfs

Observations made over the last forty years or so have shown that Sirius B is by no means unique. Procyon also has a tiny white companion, but this and the one associated with Sirius are the only two in the sun's vicinity. An ever growing number is being discovered farther afield, but their low luminosities, like those of the red dwarfs, make identification difficult. These small, hot stars, referred to as "white dwarfs," now number well over a thousand, thanks largely to the work of Luyten. Some of the very small ones have fantastically high densities. Smallest of all is LP 357–186, an object discovered by Luyten in 1962. He described it as being probably 25,000 times fainter than the sun and scarcely much larger than 1,000 miles across, or about half the diameter of the moon. Its estimated density has the outstanding value of about 100 million times that of water.

The remarkable characteristics of the white dwarfs provoke a leading question. How are stars with such high densities able to shine? The atomic nuclei and electrons in their central regions must be so tightly packed as to have

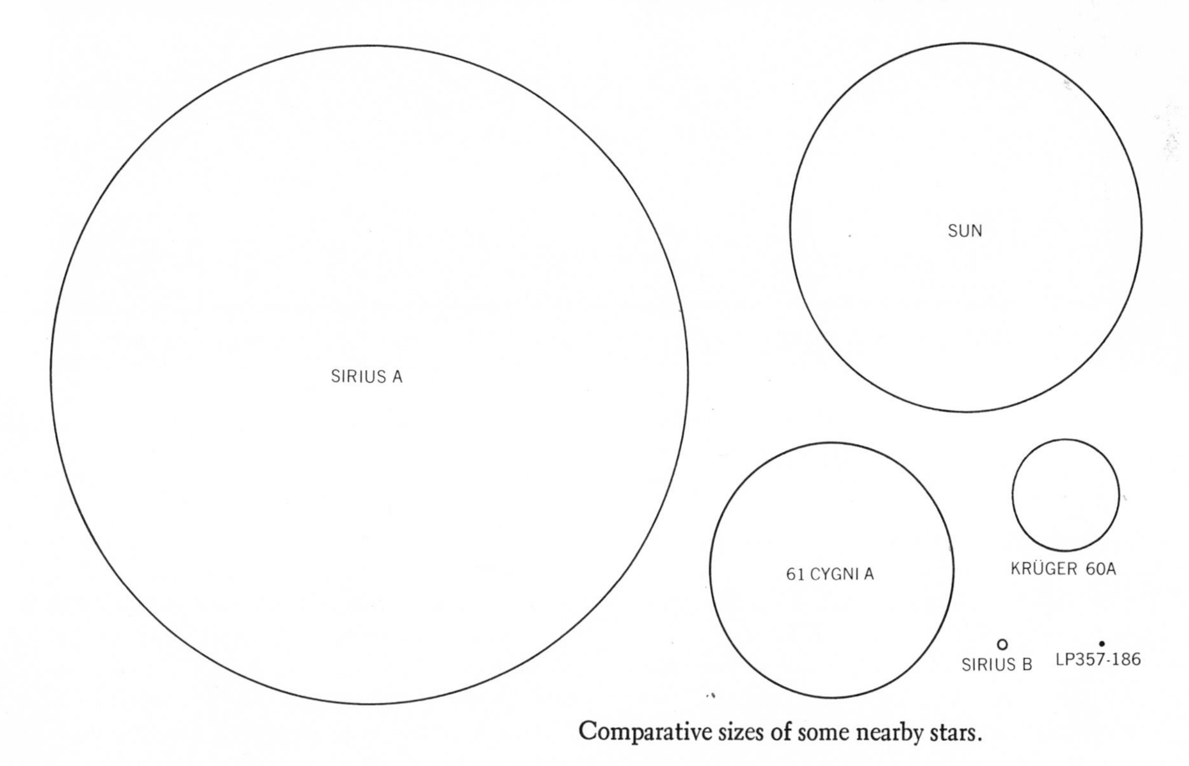

Comparative sizes of some nearby stars.

little or no freedom of movement. Theory suggests that the material has become "degenerate," or no longer capable of obeying the ordinary gas laws. The star cannot contract any more, has used up most of its internal resources of hydrogen and helium, and is deprived of nuclear energy. This condition represents, it seems, a final stage in stellar evolution. Even so, the surface area is so small that the immense store of internal energy must leak away very slowly. So once a star becomes a white dwarf it may continue to shine for billions of years.

THE TWENTY NEAREST STARS

Name	*Apparent visual magnitude*	*Spectrum*	*Distance (light-years)*	*Remarks*
Proxima Centauri	10.7	M5e	4.3	The nearest star. Flare star
Alpha Centauri A	0.01	G2	4.3	Visual binary
Alpha Centauri B	1.4	K1	4.3	
Barnard's Star	9.54	M5	6.0	Large proper motion
Wolf 359	13.5	M6e	8.1	
Lalande 21185	7.47	M2	8.2	Binary
Sirius A	−1.42	A1	8.7	Brightest star in the night sky
Sirius B	8.5	wd	8.7	White dwarf
Ross 154	10.6	M4.5e	9.3	Flare star
UV Ceti A	12.5	M6e	9.8	
UV Ceti B	13.0	M6e	9.8	Flare star
Ross 248	12.24	M5.5e	10.3	
Epsilon Eridani	3.73	K2	10.8	
Luyten 789–6	12.58	M5.5e	11.1	
Ross 128	11.13	M5	11.1	
61 Cygni A	5.19	K5	11.2	Visual binary. Small companion
61 Cygni B	6.02	K7	11.2	
Procyon A	0.38	F5	11.3	
Procyon B	10.7	wd	11.3	White dwarf
Epsilon Indi	4.7	K3	11.4	

CHAPTER V

FROM DWARFS TO SUPERGIANTS

So far our brief description of the nearer stars has been restricted to those within a radius of 13 light-years. Compared with them the sun is well above average in size, mass, luminosity, and temperature. Our next step is to consider the sun in relation to a much larger number of stars. Here we meet with a difficulty. The list of red and white dwarfs is bound to thin out with increasing distance because these stars become more and more difficult to detect. Highly luminous stars, on the other hand, draw attention to themselves by their very brightness. We cannot claim that the observed sample is a fair one, but to what extent it is unfair, future research must decide. We can, at least, be on our guard against giving undue significance to exceptional stars. A population should not be judged by its giants alone.

Stellar Classification

The twenty brightest stars, although roughly equal in apparent brightness, have a large range in distance. For some of them the method of trigonometrical parallaxes becomes no more than a rough guide to distance. For others, more remote, it is not even effective. Reliance has then to be placed on a method known as spectroscopic parallaxes. In this, stars of certain spectral types can be arranged in order of absolute magnitude by comparing the intensities of certain pairs of absorption lines in their spectra. The technique was first devised in 1914 by W. S. Adams and A. Kohlschütter at the Mount Wilson Observatory and has since been refined to estimate distances up to several thousand light-years from the earth. More recently

THE TWENTY BRIGHTEST STARS

Name	*Apparent visual magnitude*	*Spectrum*	*Parallax (seconds of arc)*	*Distance (light-years)*	*Remarks*
Sirius	−1.5	A1	0.375	8.7	White dwarf companion
Canopus	−0.7	F0	0.011	300	Supergiant
Alpha Centauri	−0.3	G2	0.751	4.3	Binary. Companion Proxima
Vega	0.1	A0	0.123	26	Optical companion
Capella	0.1	G0	0.073	45	Binary
Arcturus	0.2	K2	0.090	36	
Rigel	0.2	B8	0.004	850	Binary with binary companion
Procyon	0.3	F5	0.288	11	White dwarf companion
Achernar	0.5	B5	0.023	75	
Beta Centauri	0.6	B1	0.011	300	
Altair	0.8	A7	0.198	16	Optical companion
Betelgeuse	(0.9)	M2	0.005	650	Supergiant, variable. Companion
Alpha Crucis	0.8	B1	0.012	270	Pair of spectroscopic binaries
Aldebaran	(0.8)	K5	0.048	65	Variable. Giant
Antares	(1.0)	M1	0.008	400	Supergiant, variable. Companion
Pollux	1.1	K0	0.093	35	Optical companion
Spica	1.1	B2	0.021	220	Binary
Fomalhaut	1.2	A3	0.144	23	
Deneb	1.3	A2	0.002	1,500	Supergiant
Beta Crucis	(1.3)	B0	0.009	370	Slightly variable

O. C. Wilson and M. K. V. Bappu have introduced another method which involves measuring the widths of certain emission features in the H and K lines. This applies to a substantial fraction of stars out to about 2,000 light-years.

The luminosities of many of the brighter stars greatly surpass those of the sun and its four bright neighbors. The latter appear conspicuous in the sky only because they are comparatively near. At the top of the luminosity scale are intense beacons such as Rigel, Deneb, and Canopus, each equivalent to the light of many thousand suns. Rigel, however, is a spectroscopic binary, but either one or both stars must be highly luminous. Rigel also has a fairly faint companion in the form of another binary, so the system consists of at least four stars. Deneb has an apparent visual magnitude of 1.26, but because of its relatively great distance (about 1,500 light-years) it must be at least 50,000 times more luminous than the sun. Its A2-type spectrum and bluish-white color indicate that its surface temperature is about 10,000 degrees centigrade and a square foot of its surface emits far more light than the same area on the sun. But temperature alone cannot account for its high luminosity: if it is a single star its diameter will be roughly 50 times that of the

Stars and nebulosities in Orion photographed in blue light with an exposure of 4 hours. Although Rigel (bottom right) and Betelgeuse (top left) are both first-magnitude stars, their marked difference in color produces a striking difference in image size.
—ROYAL ASTRONOMICAL SOCIETY

sun. Canopus, rivaled in brightness only by the sun and Sirius, is an F0-type star with a surface temperature of about 8,000 degrees centigrade. It appears to be a single object and is believed to be about 13,000 times more luminous than the sun.

Supergiants Hot, highly luminous stars like Deneb and Canopus are known as blue supergiants and yellow-white supergiants respectively. They are the exception rather than the rule, at least in the sun's vicinity, and are certainly not typical members of the local community. Much the same is true of several other stars, which although highly luminous, are reddish in color and therefore relatively cool. These are the so-called red supergiants. Their high luminosities, coupled with low surface temperatures, dictate immense sizes, while their M-type spectra, although similar to those of the red dwarfs, carry telltale signs that indicate extremely low densities and therefore very diffuse atmospheres. The pressures in the lower atmospheric layers are probably of the order of one millionth that of normal atmospheric pressure at sea level on the earth.

A well-known red supergiant is Betelgeuse, a star of spectral type M2 whose color stands out in marked contrast to the bluish white brilliance of Rigel and other hot B-type stars in Orion. The effect is even more marked on small-scale photographs of Orion taken with blue-sensitive plates. About twenty years ago the estimated distance of Betelgeuse was about 200 light-years, but recent improvements in the distance scale have increased this to 650 light-years. Its luminosity has accordingly shot up to about 13,000 times that of the sun. This means that it must be an enormous star, for the surface, being low in temperature (3,500 degrees centigrade) emits comparatively little light per unit area. Fortunately its diameter can be determined in a fairly direct way.

With the aid of an instrument known as the interferometer it has been possible to measure not only the apparent diameter of Betelgeuse but also the diameters of a number of other supergiant and giant stars. Betelgeuse is a semiregular variable—its brightness (and therefore its luminosity) varies in a somewhat

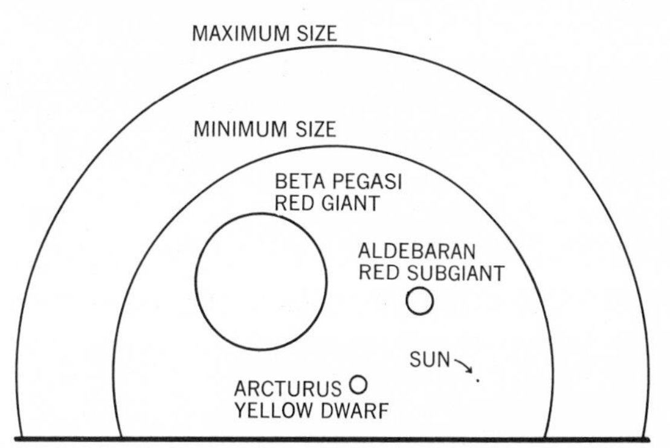

The outstanding size of Betelgeuse.

erratic manner. The magnitude limits are 0.4 and 1.3, with main brightness fluctuations once every 140 to 300 days. The interferometer reveals how this is brought about—the star pulsates, or waxes and wanes in size and therefore in surface area. Even more remarkable, the diameter is found to vary between 300 to 420 times that of the sun. Betelgeuse, with an average diameter of about 300 million miles, could easily swallow up the orbit of Mars.

The only other bright-red supergiant in the sky is Antares, or Alpha Scorpii. It is about 400 light-years away and slightly variable in luminosity. The diameter, as revealed by the interferometer, is about 285 times that of the sun. Two other well-known, although fainter, red supergiants are Ras Algethi (Alpha Herculis) and Mira (Omicron Ceti). They undergo irregular changes in luminosity in cycles of several months, and along with Antares, are associated with faint but hot companions.

Giants Since blue supergiants on the one hand and red dwarfs on the other represent two extremes in luminosity, the luminosities of the majority of stars within a distance of about 2,000 light-years can be expected to lie somewhere in-between. This is certainly true of the twenty brightest stars. All of them are more luminous than the sun and only six are cooler. Of the latter, two are the red supergiants Betelgeuse and Antares. The others, Aldebaran, Arcturus, Pollux, and Capella (a binary), are larger than the sun. These four, large, luminous, and moderate in surface temperature, are called "giants."

The Main Sequence The list of the twenty brightest stars includes three blue super-

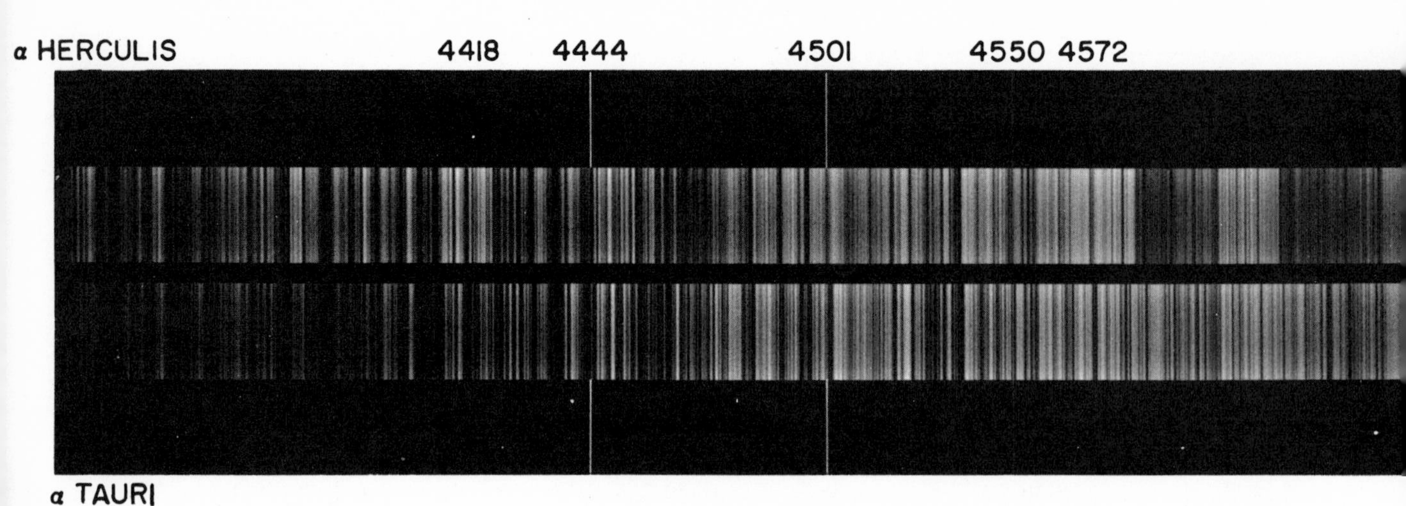

Spectrum of Alpha Herculis, an M5 supergiant with a surface temperature of about 2,800 degrees centigrade, and of Aldebaran, a K5 giant with a surface temperature of about 3,800 degrees centigrade. The general sharpness of the dark lines is characteristic of the spectra of stars with extremely diffuse atmospheres.
—ROYAL ASTRONOMICAL SOCIETY

giants, two red supergiants and four giants. The remainder, when arranged in order of decreasing temperature, luminosity, and size, form part of a definite distribution known as the main sequence or dwarf sequence. Being a sequence of decreasing temperature it is also one of changing spectral type. Type B is represented by Beta Crucis, Spica, Alpha Crucis, Achernar, and Regulus; type A by Vega, Sirius, Altair, and Fomalhaut; type F by Procyon; and type G by Alpha Centauri and the sun. The sequence also includes K-, M-, and N-type dwarfs, the majority of which are comparatively faint but near neighbors of the sun.

The significance of this sequence increases when we consider another stellar attribute—mass. In 1924 A. S. Eddington found from purely theoretical considerations that there was a close correlation between the mass and the luminosity of a star. The luminosity concerned *all* the radiation emitted by a star, not just those parts to which our eyes or photographic plates are sensitive, so it is referred to as the "absolute luminosity" or "bolometric luminosity." According to Eddington, absolute bolometric magnitude was roughly proportional to the logarithm of mass. To put it another way, the bolometric luminosity increased rapidly with small increases in mass. For example, ratios of 1 to 2 and of 1 to 10 in the masses of two stars corresponded respectively to ratios of roughly 1 to 8 and 1 to 1,000 in bolometric luminosities. In Eddington's time there was no detailed knowledge of thermonuclear reactions, so his discovery was really no more than a brilliant guess. But once astronomers knew how the stars shone, the mass-luminosity relation became of fundamental importance in studies of stellar interiors and stellar evolution.

RANGE OF STELLAR MASS Eddington's theoretical work also led him to conclude that stars over 100 times as massive as the sun would be blown apart by the pressure of radiation coming from their interiors. Similar studies by Schwarzschild and Harm have also shown that stars with masses greater than 60 suns would start pulsating as soon as thermonuclear processes took over as the main energy source. They also showed that the pulsations would build up in amplitude in less than half a million years and reach a strength great enough to disrupt the star.

So far the observational data has not led to any modification of these limits, but that concerning the masses of large stars is both scanty and generally unreliable. Zeta Scorpii, a hot B1 supergiant, is sometimes referred to as the most massive star, but since it is not known to be a

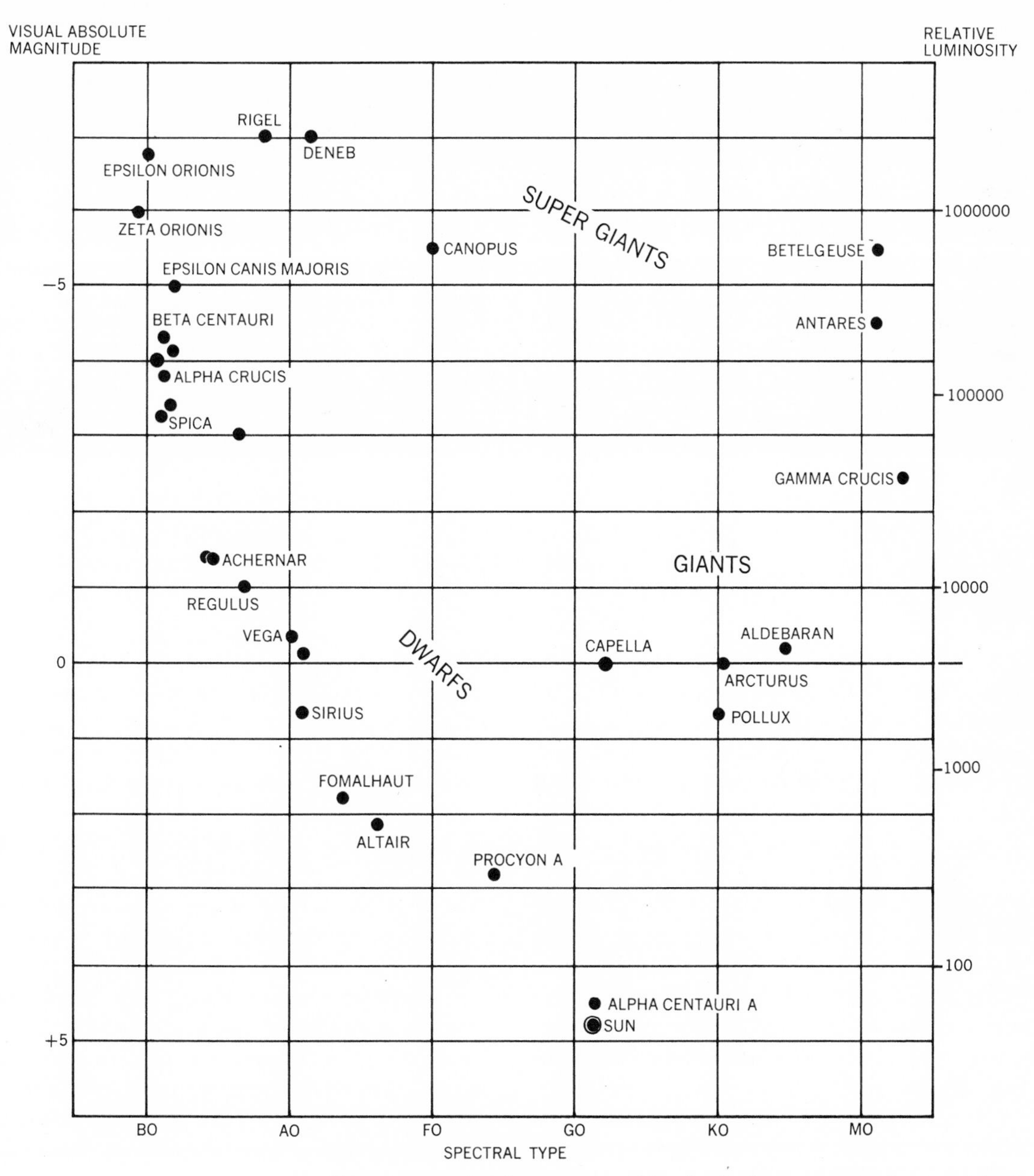

The twenty brightest stars. Visual absolute magnitude is plotted against spectral type.

binary its mass cannot be determined directly. The estimated value, about 94 solar masses, is based solely on the exceptionally high absolute bolometric magnitude of −11.8. Another very massive star is Plaskett's star, or HD 47129. Actually it is a binary whose components are thought to be 40 and 60 times more massive than the sun. At the lower end of the mass scale are the red dwarfs, one of which, Ross 614B, is considered to have a mass of 0.08 sun, the smallest yet found for a visible star. Even so, it is still 80 times as massive as the planet Jupiter. If an embryo star had a mass much lower than this it would probably never become visible, for gravitational contraction would be unable to bring about a temperature high enough to trigger off thermonuclear reactions.

DETERMINATION OF STELLAR MASSES Nearby binary stars whose orbits and periods of revolution can be determined without much difficulty, provide the only sure way of obtaining knowledge of individual stellar masses. Unfortunately, stars of this type are relatively few in number, but there is now no doubt that the mass-luminosity relationship is reasonably reliable over the great range of main sequence stars. The sequence represents a continuous distribution in temperature, size, luminosity, and mass, but as we shall see later, the most important of these is mass.

THE H-R DIAGRAM The general trends of these stellar attributes among nearby stars is well shown when their absolute magnitudes (or luminosities) are plotted against their special types (or temperatures). Two main series of stars emerge. Running diagonally across the diagram from top left to bottom right is the main sequence. Its members diminish rapidly in luminosity with advancing spectral type and form, so to speak, the backbone of the stellar community. About 85 percent of nearby stars are thought to lie on the main sequence. To the upper right are the giants. These stars are mainly of spectral types G, K, and M, and their luminosities, higher than that of the sun, have no special relation to spectral type. At the top of the diagram is the comparatively small group of supergiants, while at the bottom left are the white dwarfs.

This diagram is known as the Hertzsprung-

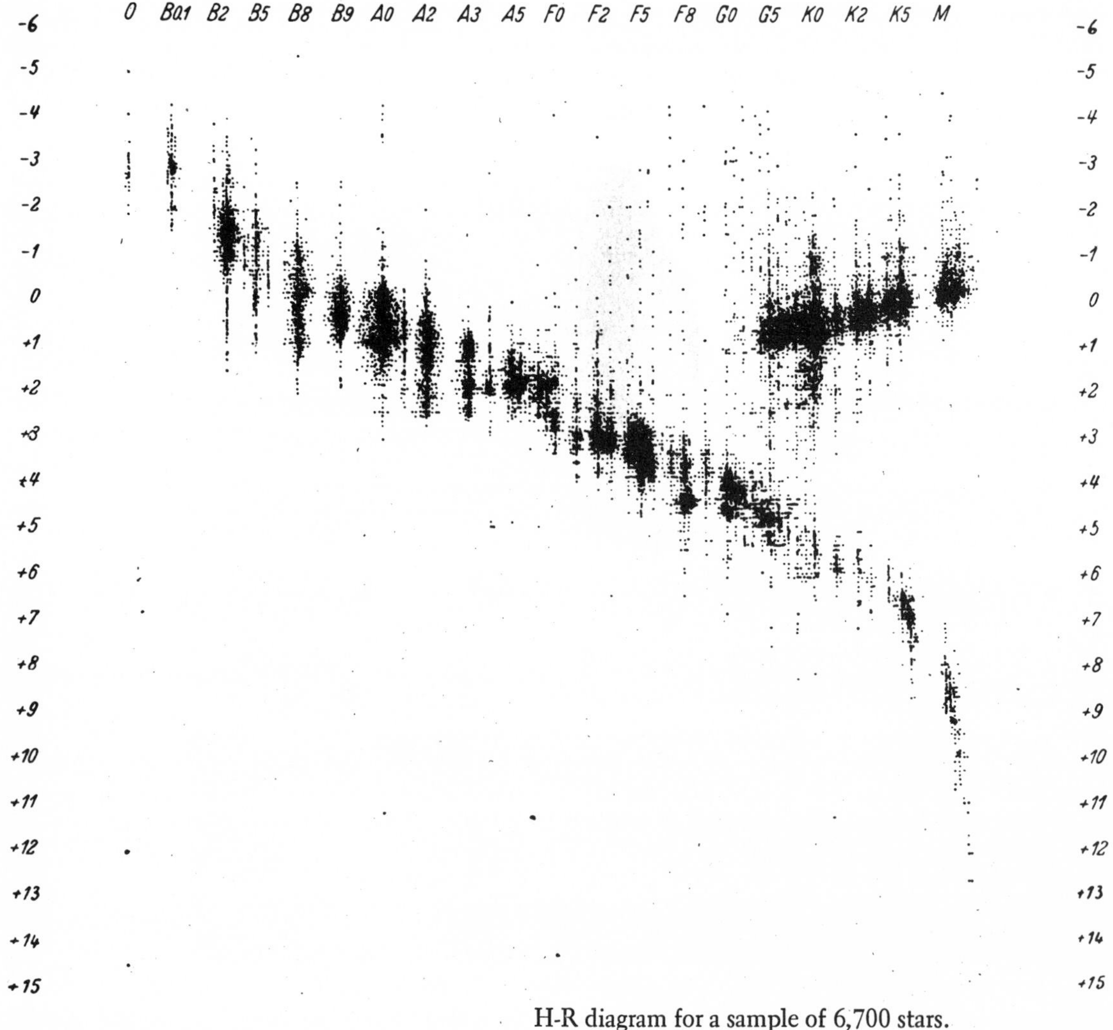

H-R diagram for a sample of 6,700 stars.
—W. GYLLENBERG, LUND OBSERVATORY, SWEDEN

Russell diagram, or H-R diagram, after E. Hertzsprung of Denmark and H. N. Russell of Princeton University. In 1905 Hertzsprung first pointed to the existence of two distinct types of red stars which he named giants and dwarfs. Russell's independent investigations, based on data for a larger number of stars, led him in 1913 to produce the first H-R diagram. Discoveries in the early 1920's led to the introduction of the classes of supergiants and white dwarfs, also of a new spectral type, S, while more recent studies have given rise to further additions in the form of subgroups.

Carbon Stars The small class of N-type stars does not belong to the main sequence. Its members are all red giants, and in a few cases, supergiants, a characteristic they share with the stars of another small class known as spectral type R. The spectra of both types have strong bands due to absorption by diatomic carbon, hydrocarbon, and cyanogen, as well as band heads due to several different carbon isotopes. For this reason they are referred to as "carbon stars," but how their atmospheres have managed to become so rich in carbon remains something of a mystery. Many of them are long-period variables; that is, they fluctuate in brightness but require months to go through a complete cycle of light variation. Their spectral types are then recorded as Ne and Re since their spectra invariably show bright emission lines during the maximum phases of their light cycles.

In 1922 astronomers approved the addition of a spectral type S to the lower end of the main sequence, but its representatives, very red in color and low in surface temperature, are now known to be giants. Their spectra are characterized by prominent absorption bands due to zirconium oxide, but it is not yet certain whether they form a continuous sequence with type M. Many of them, like R Andromedae and R Geminorum, are long-period variables and have bright emission lines in their spectra.

Subdwarfs In the 1930's W. S. Adams discovered a few stars which appeared to represent yet another stellar group or class. Christened "intermediate white dwarfs," and later, "subdwarfs," they had, according to their color indices, higher temperatures than those of main

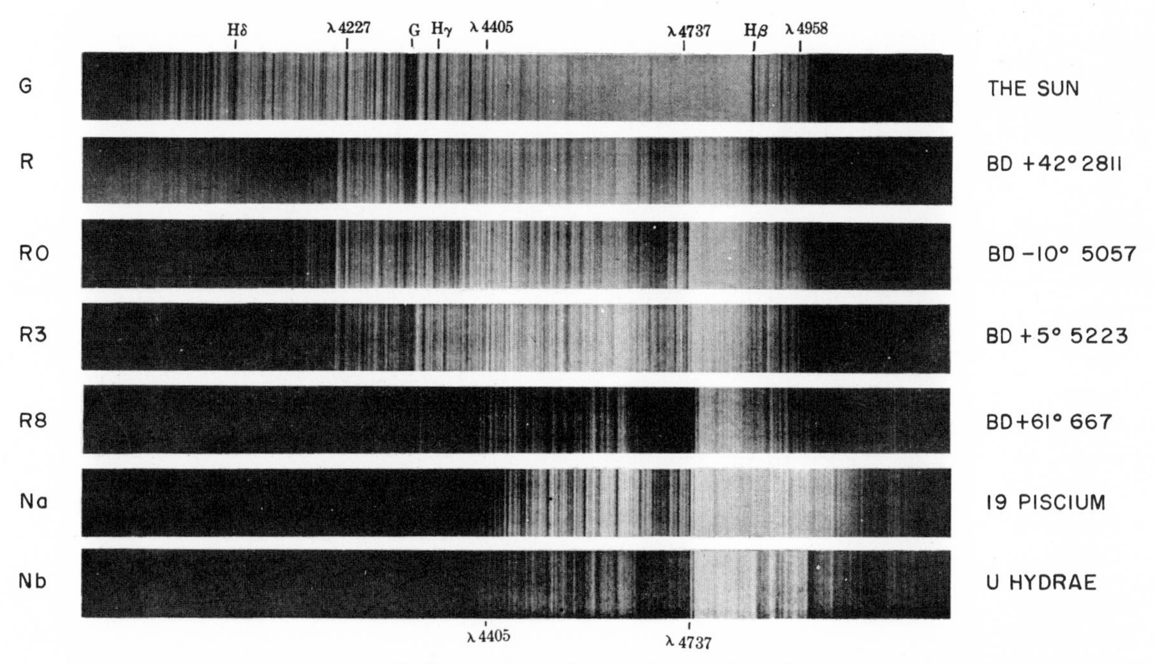

Stellar spectra of types R and N with solar spectrum for comparison.
—Royal Astronomical Society

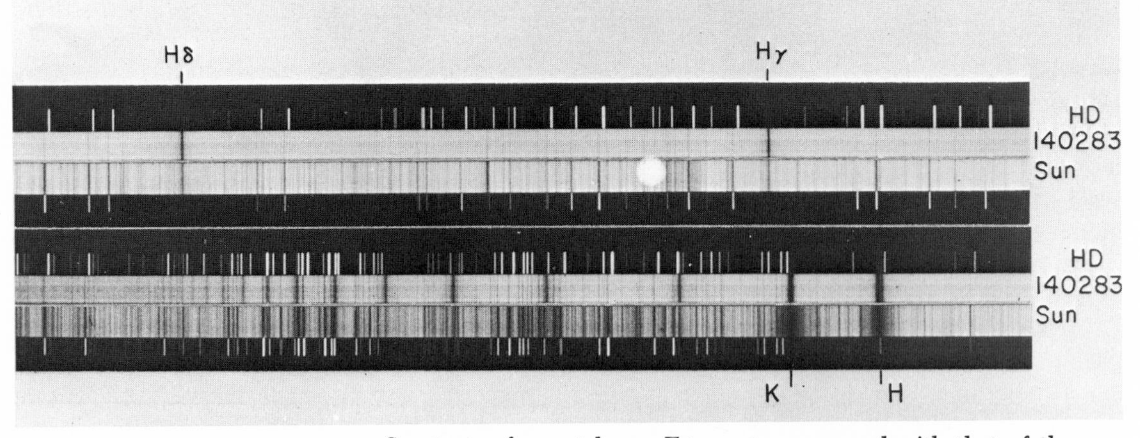

Spectrum of a metal-poor F-type star compared with that of the sun, a metal-rich star.

sequence stars of the same spectral type. Because of this, their distribution on the H-R diagram ran parallel to but slightly below the main sequence. At first, certain peculiarities in their spectra suggested that they had densities greater than those of main-sequence dwarfs. Many were also abnormally bright in the ultraviolet, and this was at first interpreted as a sign of high temperature.

It is now apparent that the atmospheres of these stars are deficient in heavy elements or metals. As a result they are more transparent to ultraviolet radiation from their deeper levels than are the atmospheres of metal-rich stars like the sun. When this and other features are considered, the majority of subdwarfs become "normal" in the sense that their distances from the main sequence are not significant.

The subdwarfs also have another curious trait—they all travel in one direction with high velocities relative to the sun. The direction is opposite to that of the sun's orbital motion about the center of the galaxy, so they must be outsiders passing by or through the solar neighborhood. This, coupled with the fact that their atmospheres are metal-poor, signifies that they cannot be generically related to the sun. We shall see later that they are members of a comparatively old population of stellar objects known as Population II stars.

Stellar Evolution

An H-R diagram for the nearer stars is no more than a kind of snapshot of their present individual states with regard to luminosity and temperature. It would be wrong to think that the various groups represent different kinds or species of star. Every star is at a particular stage of its long life-history, and the stage reached varies from star to star. It is reasonable to regard the various groups as signposts to the direction of the general evolutionary path or paths of the stellar community. The main problem has been to determine which way the signposts pointed and then to get them in the "right" order.

Toward the end of the last century the English astronomer J. Norman Lockyer put forward a theory of stellar evolution which required a two-branch classification. He sug-

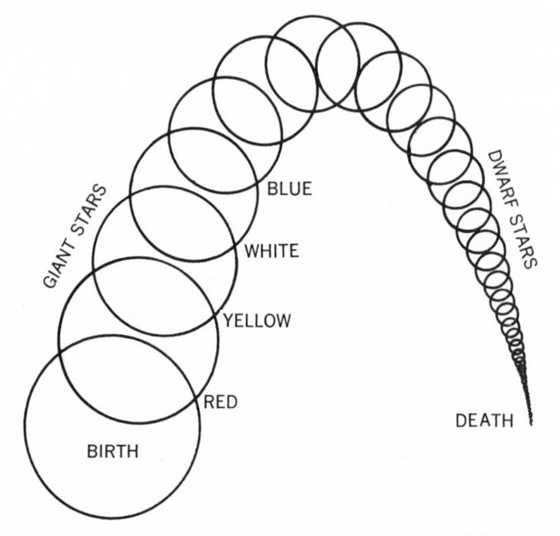

Life history of a star according to J. Norman Lockyer: the diameter of a star at birth was about one hundred times its diameter at extinction.

gested that a star was born from a seething swarm of solid metallic particles or meteorites. As the swarm slowly condensed, collisions between the particles became more frequent and the temperature rose. A giant M-type star emerged as the entire mass vaporized. The star continued to contract and increase in temperature through the range of spectral types K, G, F, A, to B. It then decreased in temperature but continued to contract, and after going through types B, A, F, G, K, and M, ended as a meteoritic lump of solid matter devoid of heat and light. Stars on the branch of increasing temperature were giants compared with those on the branch of decreasing temperature.

RUSSELL'S TWO-BRANCH THEORY Russell introduced a similar two-branch theory some twenty-five years later. The existence of giant and dwarf stars was then pretty well established and their differences in luminosity stood out clearly on the H-R diagram. In Russell's opinion a star started off as a wholly gaseous red giant and then shrank under its own gravitation to become hotter and more dense. It moved from right to left across the top of the H-R diagram and then slid down the dwarf sequence toward final extinction. Only the most massive stars reached the lofty heights of types O and B. Those less massive moved over into the dwarf sequence at types A, F, G, or K, while one with a mass as low as the sun's did not even reach type K. In making mass the main evolutionary factor, Russell anticipated, although only in a nonqualitative way, both the mass-luminosity relationship and present ideas about stellar evolution.

MODERN THEORIES These and other early ideas were completely overturned when it was realized that stellar energies were largely maintained by thermonuclear processes. In 1936 H. Bethe in the United States and independently C. F. von Weizsäcker in Germany proposed the carbon cycle as the source of solar energy. In the same year, C. Critchfield in the United States found that the proton-proton cycle could also be at work inside the sun. In the stellar field A. R. Sandage, M. Schwarzschild, Fred Hoyle, and others have made detailed calculations of the structure of stellar interiors, while an immense amount of new information has been gathered on the observational front. As a result astronomers now have a good idea not only of how individual stars grow and decay but also of the way they vary in chemical composition from one generation to another. These changes come under the general heading of stellar evolution, but the term "evolution" is generally used in a restricted sense to describe the life-histories of individual stars.

A star is now thought to begin its career as a cloud of interstellar gas and dust. As the cloud condenses, gravitational energy is converted into heat energy and the temperature increases. This continues until the temperature at the center is a few million degrees, at which time thermonuclear reactions begin. Hydrogen is converted into helium by the proton-proton reaction and the star, now full-fledged, takes its place on the main sequence. Just where it joins the sequence depends on its mass and the amount of hydrogen available as nuclear fuel.

EVOLUTIONARY SIGNIFICANCE OF MAIN SEQUENCE Once a star has settled on the main sequence, nuclear reactions supply all the radiated energy. The reactions are restricted to the

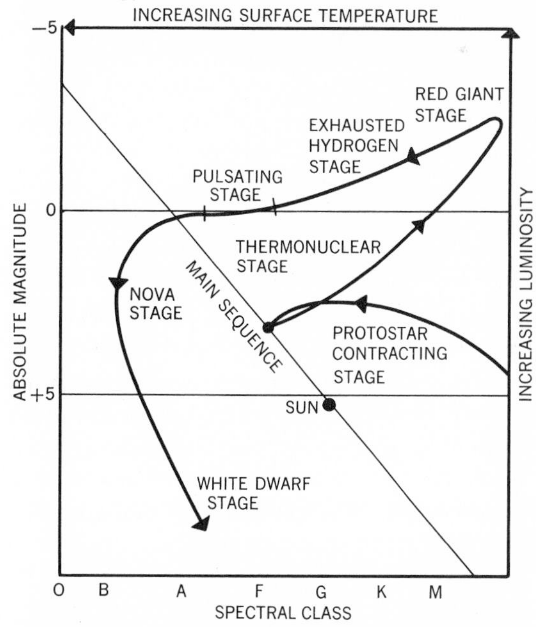

Probable evolutionary path of a star with an original mass two to three times that of the sun. After O. Struve.

material in the core, outside of which little mixing takes place. So prodigious is the store of hydrogen that the energy-balance, or "steady state," lasts a long time. A star spends most of its life on the main sequence, but the length of time it stays there depends on its mass. A highly luminous star with large mass will consume its nuclear fuel at a rate faster than that of a star of low luminosity and small mass.

Stars on the main sequence are at an early stage of their evolution. The stage is also a long one, which explains why there are more stars on the sequence than elsewhere on the H-R diagram. A star about one third the mass of the sun joins the sequence at spectral type M and stays there for many billion years. Stars similar in mass to the sun enter the sequence at type G and last for more than 10 billion years, but those of about 20 solar masses will join the sequence at type O and stay there for a few million years.

When the hydrogen in the central regions of a star is almost exhausted, the core begins to contract and the burning zone spreads slowly outwards to where the hydrogen is still plentiful. As a result the whole star expands and its outer regions cool, but the increase in surface area is large enough to bring about an increase in luminosity, despite the fall in temperature. The star therefore moves upward and slightly to the right of its main-sequence position. This trend is maintained while the expansion continues, and the star eventually reaches giant or supergiant status.

Stars similar in mass to the sun grow from the dwarf to giant states in a few million years. During the change the helium core becomes hot and dense enough to bring about thermonuclear reactions in helium and the consequent formation of carbon and other elements heavier than hydrogen. So while the temperature of the surface of a giant star is comparatively low, that of its central region must be extremely high.

Helium burning presumably keeps a star in the giant or supergiant state for several million years. But when most of the helium is used up the core again contracts, with consequent rise in temperature. This may then bring about further nuclear activity in which carbon acts as the nuclear fuel and forms neon and sodium, together with hydrogen and helium, but the reactions are too many and too complex for description here. For many stars this fairly late stage in their evolution is attended by instability. They pulsate rhythmically as Cepheid-type variables. Their mass, and therefore their luminosity, determines the period of light variation: large mass is associated with a short period and vice versa.

If a star has a very large mass it may become so unstable that it explodes. The explosion practically turns it inside out, and the flood of energy can raise the luminosity to that of a billion suns. But by the time the energy reaches the earth we see at most no more than a "super new star," or "supernova." A star appears to come from nowhere, rise rapidly in brilliance, and then fade slowly into obscurity. During the explosion the star scatters most, if not all, of its mass in a rapidly expanding cloud of gas. In this way heavy elements born within a star are distributed over a vast region of space and mixed with the clouds of interstellar hydrogen already there. So if the clouds subsequently condensed into stars, the new stellar generation would contain a larger admixture of metals than did the old one. Differences of this kind *are* found in stellar populations, so they probably arise from corresponding differences in the star-building material itself.

The clue to the fate of stars that survive the instability stage is provided by the white dwarfs. As we saw earlier, these stars are hot, faint, small, and extremely compact, but their nuclear fires are figuratively reduced to ashes. Lying to the bottom left of the diagram they represent stars that have undergone so large a decline in size and luminosity as to drop below the main sequence. Recent studies by J. Greenstein and O. Eggen show that the atmospheres of some white dwarfs are composed largely of helium, while those of others consist mainly of carbon. This in turn suggests that some stars can continue their thermonuclear processes well beyond the average span.

The observational data certainly seems to support the theoretical picture of the majority of stars ending as white dwarfs. This means that stars shed one half or more of their original mass somewhere along the evolutionary path, for according to a theory of the structure of white dwarfs developed by S. Chandrasekhar in 1935, a star can become a white dwarf only if its mass is less than 1.44 the mass of the sun. Supernova outbursts are too rare to be significant in this connection, and the same is also true of less violent although more frequent outbursts known as novae. Nor can the loss be laid at the door of the normal conversion of mass into radiation. A star like the sun loses mass in the form of radiation at the rate of 4 million tons a second, but the overall loss in several billion years is a mere trifle compared with the original mass. Even if the sun were made entirely of hydrogen and converted it all into helium, the loss of mass would be less than one percent of the whole.

There are now indications that the solution could lie in the red giant and supergiant phases, for Alpha Herculis, Antares, and Eta Geminorum have been found to be embedded in vast expanding shells of highly tenuous gas. Similar expanding shells may surround Betelgeuse and other red supergiants, also the red giants, although to a smaller extent. The notion receives further support from the discovery, made by F. J. Low in 1964, that Betelgeuse, Aldebaran, and Mu Cephei (a red supergiant) have faint but extensive infrared envelopes. Low found that the envelopes appeared to change in brightness, shape and size from night to night, and concluded that they were being excited by energy released from the stars that they surrounded.

It must be stressed that the modern picture of stellar evolution is generally thought to be on the right lines only for the main sequence and giant or supergiant stages. The early phase, from the time when stars are formed to their arrival near the main sequence, is still poorly understood. What might happen after the giant stage has been traced theoretically, but the observational support is still extremely patchy. The problem bristles with difficulties, as we will learn when we consider, in the next two chapters, the great variety of stars. Binary and multiple stars, once thought to be comparatively rare, now appear to be very common, while variable stars of many different kinds occur in great numbers. One problem, among others, is to assess the evolutionary significance of these and other "peculiar" members of the stellar community. In this, as in all other areas of scientific enquiry, theory is only a guide. It must not be allowed to influence the selection of data, nor can it stand aloof from the evidence of observation.

CHAPTER VI

BINARY AND MULTIPLE STARS

Although most of the brighter stars appear to be single objects to the unaided eye, at least one third are double or multiple. Of these a large number are binary systems (as distinct from optical doubles), but their components vary widely in separation and relative brightness. A few binaries, like Alpha Capricorni, Epsilon Lyrae and Theta Tauri, can be seen as pairs of stars with the naked eye. The rest have components separated by less than 1 minute of arc and can therefore be resolved, or seen as two, only with the telescope. The ease with which this can be done depends mainly on their angular separation and the aperture of the telescope. When the separation is very small, about 0.25 second of arc for example, the components can barely be resolved by a large telescope 20 inches or so in aperture.

RELATIVE ORBITS Each component of a binary system describes an ellipse around the mass-center of the system. During the course of observation, however, one component appears to travel about the other in an ellipse. The latter is only a relative orbit and is seldom orientated face-on. In most cases its plane is tilted to the plane of the sky background, so it appears foreshortened. But once the period of revolution and the shape and linear size of the true relative orbit are known, the combined masses of the two stars can be derived by applying Newton's law of gravitation. The same law also enables the relative masses of the components to be determined, but in this case the shape and size of the actual orbit of either star about the mass-center must be known.

Some binaries, with components fairly evenly matched in brightness and moderately spaced, have been assiduously observed over many decades. One of these is Gamma Virginis, whose components, each of the third magnitude, revolve about each other in a period of about 172 years. Its binary nature was discovered in 1780 by Sir William Herschel, who estimated that the two components were nearly 6 seconds of arc apart. They have completed just over one revolution since Herschel's time, and their separation and relative positions are

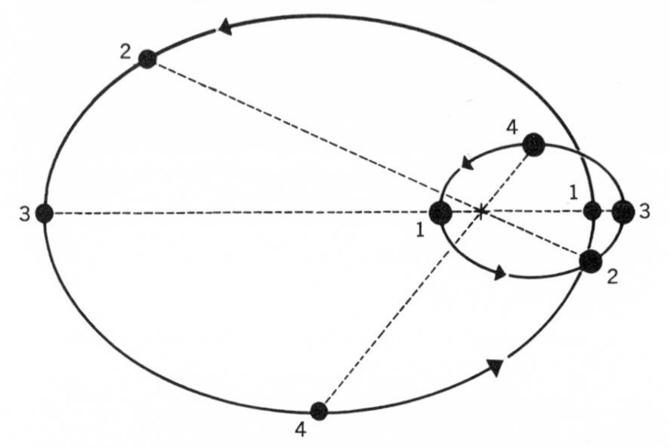

A binary system. The two stars describe their own individual orbits about the mass-center of the system. In this diagram the primary star is assumed to be three times as massive as the secondary. The mass-center divides the distance between the two stars in a 3 to 1 ratio.

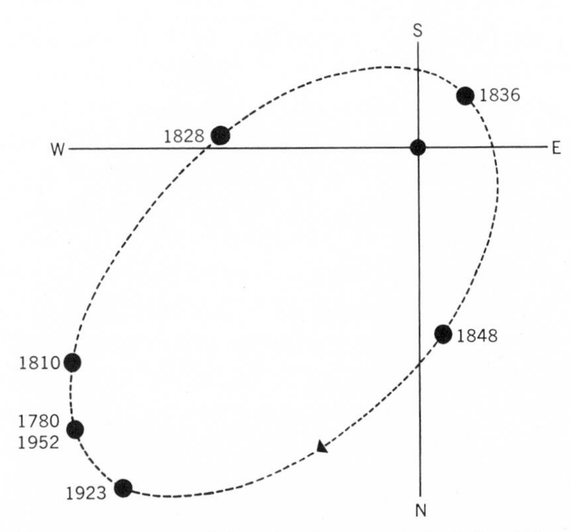

The apparent orbit of Gamma Virginis. The component stars are about equal in brightness and each is 1.3 times as massive as the sun.

now roughly the same as they were in 1780, the year of Gamma Virginis' discovery.

Other bright binaries with well-matched and well-separated components are Alpha Centauri (80 years), Gamma Leonis (400 years), Castor, or Alpha Geminorum, (400 years) and Mizar, or Zeta Ursae Majoris, (60 years). As the decades pass the components of these and similar binaries appear to swing together and apart owing to their orbital motions and the tilt of the plane of their orbits to the line of sight. In consequence they may be sometimes difficult and sometimes easy to resolve.

GENERAL PROPERTIES Binaries with periods of many years have one thing in common: their components move in fairly eccentric orbits. The components of Alpha Centauri, for example, can be as near together as 11 A.U. (slightly greater than the distance of Saturn from the sun) or as far apart as 35 A.U. (nearly the distance of Pluto from the sun). In contrast, the components of binaries with periods of a few years move in orbits which are smaller and more nearly circular.

Components approximately equal in luminosity are generally of the same or nearly the same spectral type. When they are main-sequence stars and differ in luminosity they also differ in spectral type. If the primary or more massive star is a giant, the companion may be either a giant or a dwarf. In the former case the surface temperature of the companion is usually higher than that of the primary, while in the latter it is usually about equal.

FREQUENCY OF VISUAL BINARIES In 1932 the Carnegie Institution of Washington published a catalog of 17,180 visual double stars prepared by R. G. Aitken of the Lick Observatory. They were contained within the declination limits +90 degrees and −30 degrees and indicated that there was about one visual binary for every eighteen stars. The actual frequency, however, must be considerably higher, for the number of known binaries has now grown to about 60,000. Yet accurate information about the masses of components is still comparatively scarce, due largely to the low accuracy of distance determinations. The situation is complicated by the fact that the stars of many binaries are themselves binaries, so that what appears to be a double star may be a system of three, four, and even more stars.

Spectroscopic Binaries

Binary stars whose components are too close to be resolved by the telescope can be detected with the spectroscope. As the components gyrate round each other the lines in their spectra undergo small periodic shifts due to the Doppler effect. Systems that can be recognized in this way are known as "spectroscopic binaries." The line shifts are greatest when the planes of the orbits lie in the line of sight and one component is approaching us while the other recedes. The observed magnitude is then a maximum since it represents the combined brightness of both stars. In between these positions each component in turn eclipses the other. At this point the radial velocities drop to zero and the combined brightness may be greatly reduced. Spectroscopic binaries which fluctuate in brightness owing to these periodic eclipses are called "eclipsing binaries." Usually the eclipses are only partial since the plane of motion of the two stars is seldom precisely in the line of sight.

The periods of revolution of most spectroscopic binaries range from one year or two years down to a few hours. Of course, a fairly well separated pair may be so distant that its duplicity can be detected only by the line-shifts in its spectrum. The period can then be several years, but the star is still referred to as a spectroscopic binary. When the lines are sharp and their relative shifts are comparatively easy to measure, the period can be found from the velocity curve, or the graph obtained when the radial velocities of each component are plotted against time. If the stars move in circular orbits the curve takes the form of a pair of sine waves, although usually the secondary star is too faint to give a spectrum, and only one sine wave can be drawn.

The velocity curve can also be used to determine the ratio of the masses of the component stars. But since their mean separation and the inclination of the plane of their orbital mo-

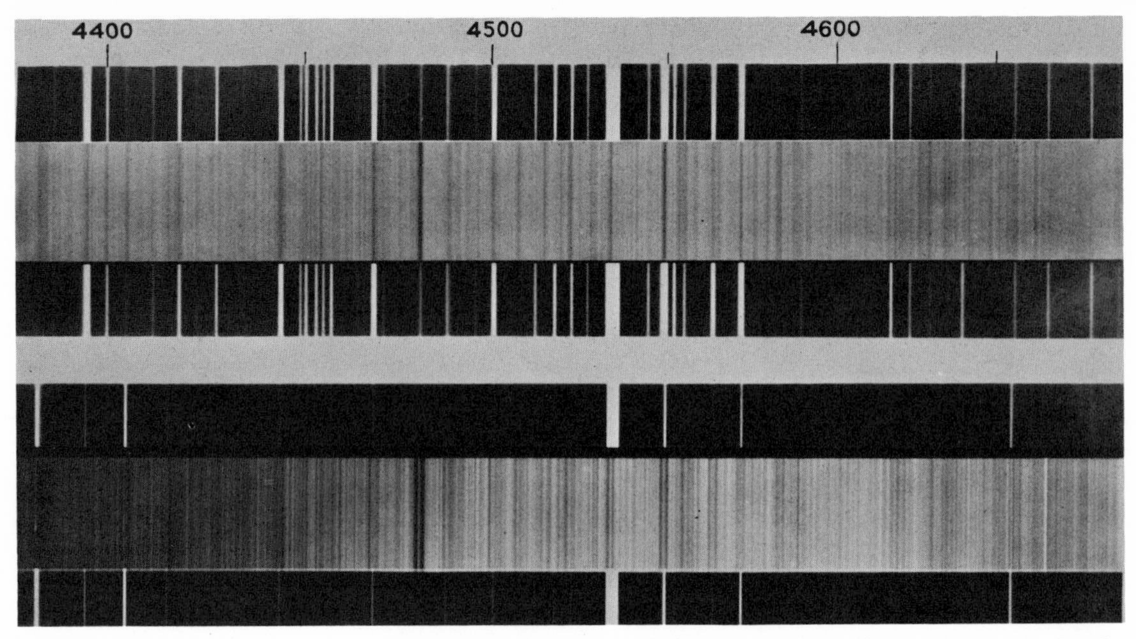

Part of the spectrum of the spectroscopic binary Mizar A. In the upper photograph the dark lines are single because the two stars are moving across the line of sight and their radial velocities are zero. In the lower photograph the lines are double because one star is moving toward us and the other is moving away. The bright lines on either side of the spectra are for comparison purposes only.

tion to the line of sight are unknown, it is impossible to find their individual masses. In this respect eclipsing binaries are more accommodating. The plane of the orbital motion of their components is almost in the line of sight and their mean separation can be derived from their orbital velocities. The latter are always constant when the orbits are circular so it is possible to determine the relative masses in the same way as with visual binaries.

The spectroscopic method shows that some 40,000 stars are double, while a large number of the well-known visual binaries are multiple systems. Of the thirty nearest stars only seventeen are single objects, whereas the other thirteen together have twenty-nine components. Thus 63 percent of a total of forty-six stars belong to binary or multiple systems. Similarly, of the thirty brightest stars in the sky, fifteen are composed of forty-one individual components. In this case the percentage of objects belonging to systems is 73 percent. But until corresponding information about a much larger sample is available it would be misleading to regard single stars as exceptional. On the other hand, astronomers would not be at all surprised if this actually turned out to be the case.

MIZAR The study of spectroscopic binaries began in 1889 when Miss A. C. Maury noticed on Harvard observatory photographs the radial-velocity effect in the spectrum of Mizar, a well-known visual binary in the handle of the Big Dipper. The line shifts indicated that Mizar A was itself a binary with components nearly equal in brightness revolving about each other in a period later shown to be 20.5 days. There is now evidence that Mizar B also is a spectroscopic binary, with a period of 182 days, and that it is probably associated with a third component orbiting the pair in 1,350 days. In addition, Alcor, a fourth-magnitude star 12 minutes of arc away from Mizar and a member of the system, is also a spectroscopic pair. So although Mizar and Alcor appear as two single stars to the unaided eye, they really represent a family of seven stars.

Among the brighter stars, Capella, Spica, and Lambda Scorpii are spectroscopic binaries, while Rigel, Acrux (Alpha Crucis), Castor, and Alpha Herculis are multiple systems. Rigel has been known to have a faint companion

(Rigel B, magnitude 6.8, distance 10 seconds of arc) ever since the time of Sir William Herschel. Yet both Rigel A and Rigel B are spectroscopic binaries with periods of 22 days and 10 days respectively. Acrux, a bright visual

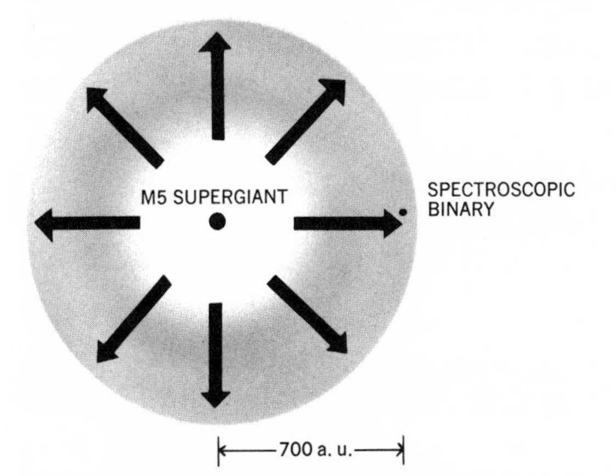

The system of Alpha Herculis according to A. J. Deutsch. The red supergiant primary star is surrounded by an expanding shell of gas that extends at least as far as the binary companion.

binary, and Castor, discovered to be a double star by Bradley in 1719, are quadruple systems. A faint spectroscopic binary fairly near Castor shares in the latter's proper motion, so the complete Castor system contains not four but six stars.

ALPHA HERCULIS In the case of Alpha Herculis, the primary, a red M5 supergiant, is a semiregular variable of magnitude limits 3.0 and 4.0. In a small telescope its orange-red brilliance stands out in beautiful contrast to the blue-green color of its sixth-magnitude companion. The latter, however, is a spectroscopic binary whose spectrum, predominantly type G0, contains traces of an A3-type spectrum. According to A. J. Deutsch, the binary lies in a diffuse expanding envelope which surrounds the primary. The envelope consists largely of hydrogen and is ionized presumably by short-wave radiation from the hot A3-type dwarf. So although the main body of Alpha itself has an estimated diameter of 580 suns, the depth of its atmosphere must exceed the distance of its binary companion. This distance is thought to be of the order of 700 A.U., or roughly 65 billion miles. A similar situation is found with Antares (although in this case the faint but hot companion appears to be a single star) and also with Eta Geminorum.

ECLIPSING BINARIES With a spectroscopic binary that is also an eclipsing binary, the extent of the eclipse and the corresponding variations in brightness depend on the relative sizes and luminosities of the components and also on the inclination of the orbital plane to the line of sight. When the light variations are plotted against time, the resulting graph is called a light curve. A precise light curve based on photoelectric measurements is a most valuable guide to the geometry and photometry of the system to which it belongs. From its shape it is possible to determine, among other things, the type of eclipse (*i.e.*, whether total, annular, or partial) and the relative brightnesses of the components. This, considered along with the orbital velocities (obtained from the relative line shifts), then gives the diameter of each star.

ALGOL One of the most famous of all eclipsing binaries is Algol, or Beta Persei, whose light changes are both conspicuous and remarkably regular. Its variability was discovered between 1667 and 1670 by G. Montanari of Bologna, but the periodic nature of the changes was first recognized by the English amateur astronomer J. Goodricke in 1783. Goodricke, only eighteen years old at the time, suggested that the periodicity might be caused "by the interposition of a large body revolving around Algol." The changes in brightness have a period, or interval between one minimum and the next, of 2 days, 20 hours, 49 minutes. For about 2 days, 11 hours Algol is of magnitude 2.3. It then gradually fades in 5 hours to reach magnitude 3.5, but in 5 hours more brightens to attain magnitude 2.3 again. The sharp dip in brightness, known as the "primary minimum," is due to the partial eclipse of the brighter star by its fainter companion. In addition, a slight decrease of about 0.05 magnitude occurs midway between two successive primary minima. This "secondary minimum" is due to the eclipse of the companion by the primary, but since the latter is much brighter than the former the loss of light is not conspicuous.

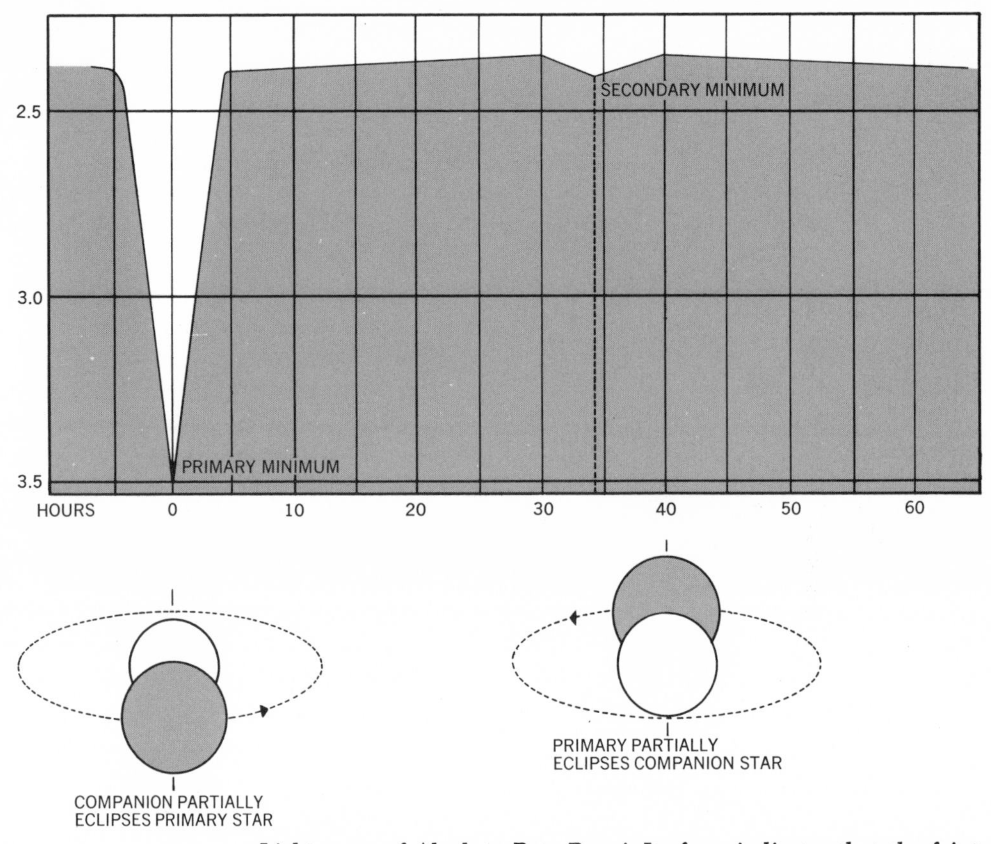

Light curve of Algol, or Beta Persei. Its form indicates that the fainter component of a binary system alternately eclipses and is eclipsed by the brighter component.

Elaborate studies of the light curve and spectrum of Algol indicate that the system is a multiple one. The two main components are separated by a distance of less than 10 million miles and revolve about each other with a period of 2.867 days. Algol A, of spectral type B8, is five times as massive as the sun and has a diameter three times the solar diameter. Algol B, of spectral type F0, is about equal to the sun in mass and slightly larger than Algol A. The short period of revolution and close proximity of the two stars suggest that their orbital periods and periods of axial rotation are identical. This theory is supported by the diffuse appearance of the spectral lines, for the effect is brought about by rapid axial rotation. It is highly probable that the components are not spherical in shape but slightly flattened or elliptical. The orbits are almost circular, but complex irregularities in the period show that the system is at least triple and probably quadruple. The F2 spectrum of the third component was first recorded in 1957 by A. S. Meltzer of the Mount Wilson and Palomar Observatories. His calculations showed that Algol C has a period of 1.8 years and is slightly larger and more massive than the sun. Yet when the effect of this third star is considered, small irregularities still remain, and these are thought to be due to the influence of at least a fourth component which, so far, has eluded direct detection.

Epsilon Aurigae Another interesting eclipsing binary of the Algol-type is Epsilon Aurigae. The main component is an F2 supergiant, but owing to its great distance it appears no brighter than a star of the third magnitude. In this case the eclipse can last nearly two years, so the eclipsing star must presumably be very large. Yet curiously enough the component shows no spectral lines: the spectral type remains F2 both during and outside of eclipse. It was thought that the component's radiation lay mainly in the infrared—that the eclipsing star was a cool, tenuous supergiant immense enough to be able to contain the entire orbit of the planet Saturn. Recent studies by Miss M. Hack, however, indicate that it is a hot but relatively small star. Its diameter is about 1/10 that of the F2 supergiant and it is surrounded by an extended gaseous ring whose inner radius is about 10 times the radius of the star itself.

Algol-Type Variables The existence of gaseous rings around the equators of the hot, luminous primary stars of several Algol-type variables is now well established. A striking example is provided by RW Tauri, whose B9 spectrum periodically shows the bright emission lines of hydrogen. The lines appear for a short time just after the hot star disappears behind its larger K0 companion and again just before it reappears. On the former occasion the lines are slightly shifted toward the red, thereby showing that the uneclipsed part of the ring is receding from us, while on the latter occasion the shift is towards the violet, showing that the uneclipsed part is approaching. It follows that the ring is rotating in the same sense as the axial rotation and orbital revolution of the hot primary.

This explanation, put forward by A. H. Joy in the early 1940's, has since been successfully applied to a number of other eclipsing binaries whose spectra show emission features. In the case of RZ Scuti and some other stars, the moving gas stream appears to surround both the hot primary and its cooler companion. O. Struve has suggested that these stars may be further examples of systems in which a cool giant star spills hydrogen-rich material into the gravitational regions controlled by a hot but small primary.

The number of stars known to be Algol-type variables is now very large. All of them appear to consist basically of a hot, luminous, main-sequence star and a cooler, less luminous sub-giant. An even larger number of eclipsing variables consist of two stars well matched in size, mass, luminosity, and temperature. Most of them are O-, B-, or A-type stars so close together that their periods are less than ten days. Their light curves show a continuous variation in brightness with decidedly smooth or rounded maxima and minima. This indicates that both components are elongated, which is what we should expect. Not only are they slightly flattened owing to their rapid axial rotations, but each star raises great tides in the other and therefore distorts it still further. Sometimes the light curve has a slight rise just

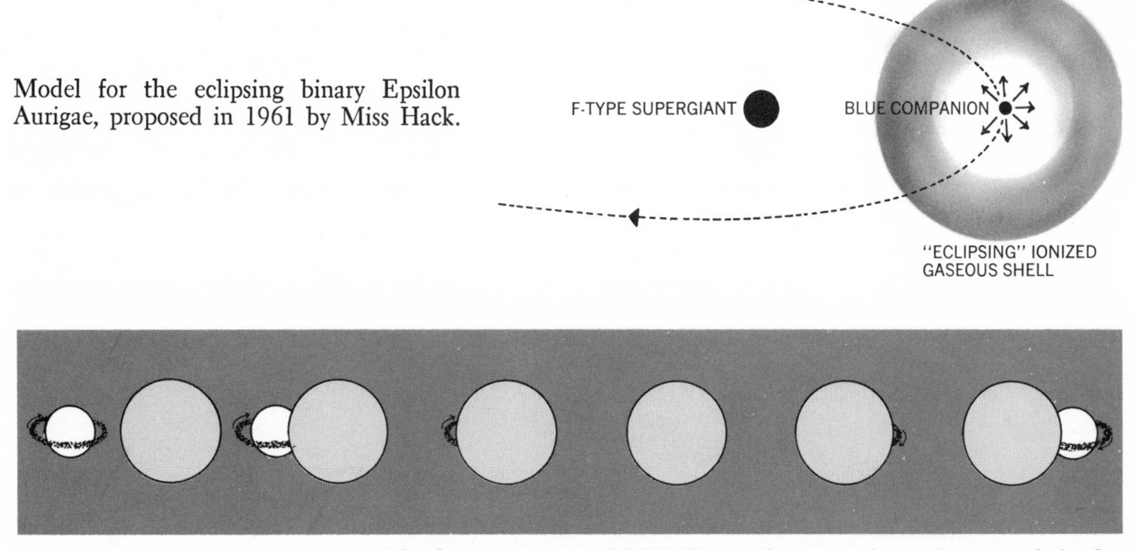

Model for the eclipsing binary Epsilon Aurigae, proposed in 1961 by Miss Hack.

The binary system of RW Tauri: the spectral peculiarities of the binary RW Tauri can be accounted for by supposing that the smaller, but hotter, component is surrounded by a luminous ring of hydrogen.

before the secondary minimum and a corresponding slight fall just after. This so-called reflection effect shows that one star is illuminating and heating the facing side of the other. Finally, slight distortions of the velocity curves and certain spectral peculiarities of some close binaries can be attributed to streams of hot gas, usually hydrogen or helium, which move from one component to the other and also envelop the whole system.

BETA LYRAE There is also evidence that the gas surrounding some close systems is rapidly expanding and carrying mass into space. The decrease in the total mass means that the periods of the components must be increasing. This is nicely illustrated in Beta Lyrae, the most thoroughly studied of all short-period eclipsing variables. Its light variations have been followed ever since they were discovered by Goodricke in 1784. The period is nearly thirteen days, and the maximum magnitude 3.4, with ranges at the primary and secondary minima of 0.97 magnitude and 0.45 magnitude respectively. According to O. Struve and others the period is increasing at the rate of 18.8 seconds a year. One component (or perhaps both) is therefore shedding its mass and expanding towards giant status.

Beta Lyrae shows a predominantly B8-type spectrum which belongs to the primary, a hot blue elliptical supergiant of unknown mass but with a diameter about 140 times that of the sun. The secondary shows no spectrum, so no velocity curve is available and the mass is unknown. Studies of the light curve suggest that the secondary is an elliptical F-type star with a surface temperature of 8,000 degrees centigrade.

Light curve of Beta Lyrae. The shape of the curve indicates that the two stars are comparable in luminosity, very close together, and elongated by the effects of tidal action.

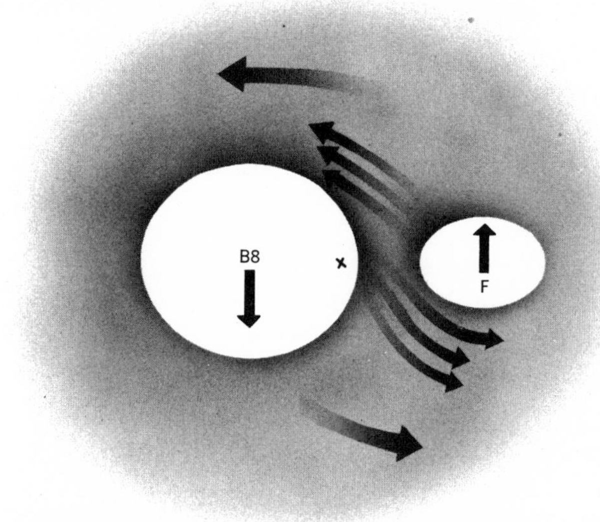

Model of Beta Lyrae, according to O. Struve. Gases stream from both stars and envelop the system in a rotating and expanding atmosphere. The cross indicates the mass-center of the system.

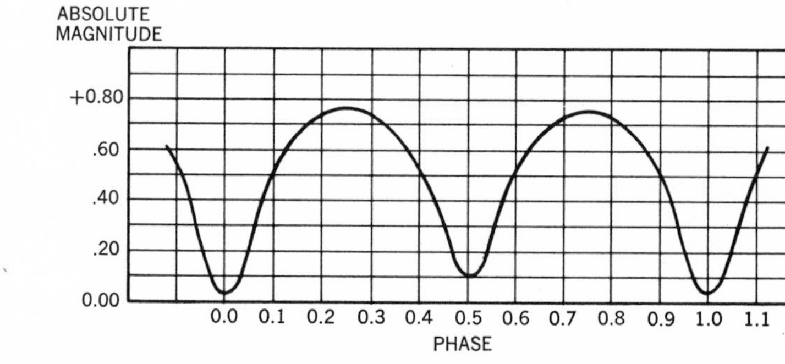

Light curve of W Ursae Majoris, based on photoelectric observations made in 1934 by C. M. Huffer at Washburn Observatory. The curve is typical of those of eclipsing binaries whose components are nearly in contact and almost equal in luminosity.

CONTACT BINARIES Many eclipsing variables, some hundreds in fact, are "contact binaries" in the sense that the two stars are so close together as to be nearly in contact. They have periods of less than 1½ days and their components are usually of spectral type G or early K. They have as their prototype W Ursae Majoris, an eclipsing binary with an unusually short period of 8 hours. The continuous change in brightness of this star, coupled with minima of nearly equal depth, shows that the component stars are elongated and practically twins in luminosity and size, although their masses are not the same.

CHAPTER VII

EXPLODING STARS

DWARF NOVAE Some variable stars are spectroscopic binaries of a noneclipsing type. Their light fluctuations are intrinsic, arising from changes in the luminosity of one or both components. Stars of this kind are characterized by a sudden increase in brightness by a factor of 10 to 100 within several hours, followed by a decline in two to three days. The flare-ups occur at intervals of weeks or months and are fairly irregular in their spacing. Stars of this kind are often called dwarf novae, but the name is misleading since the outbursts are recurrent and the stars concerned are certainly not "new." Their brightest member is SS Cygni, a faint star which occasionally reaches the eighth magnitude. One group is named after Z Camelopardalis which at times behaves in a most erratic way and then settles down at an intermediate level of brightness for several months. Another group is named after U Geminorum, a star with less frequent outbursts and a much larger range in brightness.

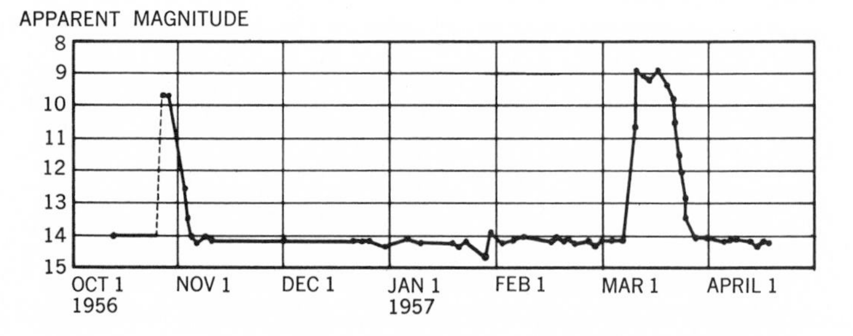

Light curve of U Geminorum for the winter season of 1956–1957, based on daily means of magnitude estimates by the American Association of Variable Star Observers. Its most striking feature is the sudden and rapid rise from magnitude 14 to magnitude 9, corresponding to a one-hundred-fold increase in brightness.

In 1943 Joy discovered that SS Cygni was a close binary with a remarkably short period of 6 hours, 38 minutes. The components are a main-sequence G5 star similar to the sun but low in luminosity for its spectral type, and a hot dwarf B-type star believed to be the seat of the light fluctuations. Joy's observations also led him to conclude that the radius of the primary star was about equal to the radius of its orbit about the mass-center of the system. U Geminorum, AE Aquarii, and several other dwarf novae have also been found to be spectroscopic binaries, so it is probable that all the rest are double. Those studied so far usually consist of a cool main-sequence star and a small but hot B-type star in close proximity. According to a model by J. A. Crawford and R. P. Kraft the cooler star ejects a stream of gas which flows around the hot one and thereby involves it in a high-temperature cloud. If this is so the dwarf novae provide yet another example of stars in the process of shedding mass through expansion. The sudden increases in brightness could be due to great flarelike explosions on the cooler components.

RECURRENT NOVAE Some stars have surges in brightness very much greater than those of the dwarf novae. Because they flare up at intervals of many years, they are called recurrent novae. Their most notable members are T Coronae Borealis (1866 and 1946), RS Ophiuchi (1898, 1933, and 1958), and T Pyxidis (1890, 1902, 1920, 1944, and 1960). About thirty years ago they were thought to be single stars, which for causes unknown occasionally ejected some of their outer material into space. In 1946, however, R. F. Sanford examined the spectral changes of T Coronae Borealis during the outburst of that year and concluded that the star was a binary with a red-giant component. His findings were subsequently confirmed by Kraft, who in 1955–1956 discovered that the other component was a hot dwarf star. The system is similar to that of SS Cygni, the more so since the radius of the red giant is nearly the same as the radius of its orbit about the mass center. More recently W. Krzeminski, working at the Lowell Observatory in Flagstaff, Arizona, has discovered that another recurrent nova, WZ Sagittae (1913 and 1946), is an eclipsing binary.

BRIGHT NOVAE The great majority of novae flare up only once. They increase in brightness by a factor of 5,000 to 100,000 within two or three days and then decrease to their former insignificant state in a matter of months and even years. Several have been prominent naked-eye objects, rising in brilliance to outshine some of the brighter stars. During this century the brightest have been Nova Persei (1901), Nova Aquilae (1918), Nova DQ Herculis (1934), Nova CP Puppis (1942), and Nova T Coronae Borealis (1946), although the last previously gave a display in 1866.

The details of the light curves of novae vary from star to star. The same is true of the complex changes in their spectra. One of the main spectral features during the outburst is the displacement of the dark absorption lines toward the violet. At maximum brightness the dark lines are accompanied by broad emission lines, but as the star fades, the former gradually disappear to leave a purely emission spectrum similar to that of a gaseous nebula. This also fades, and the spectrum finally shows broad emission lines on a faint continuous background.

The displacement of the dark lines toward the violet indicates that the star is expanding. The broad emission features can be accounted for by assuming that a shell of gas is moving outward from the star in all directions, while

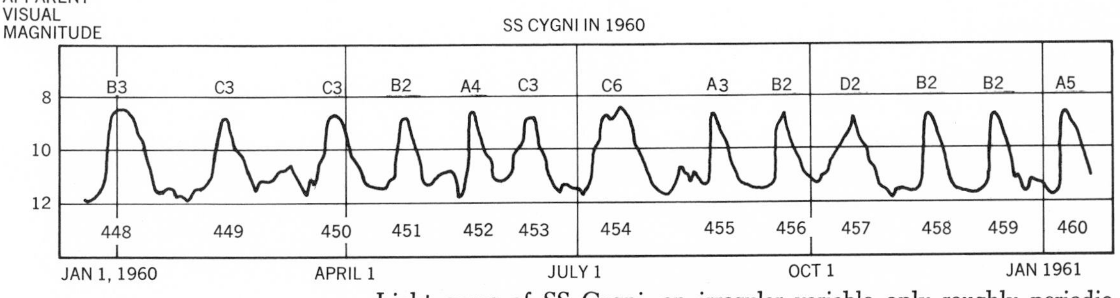

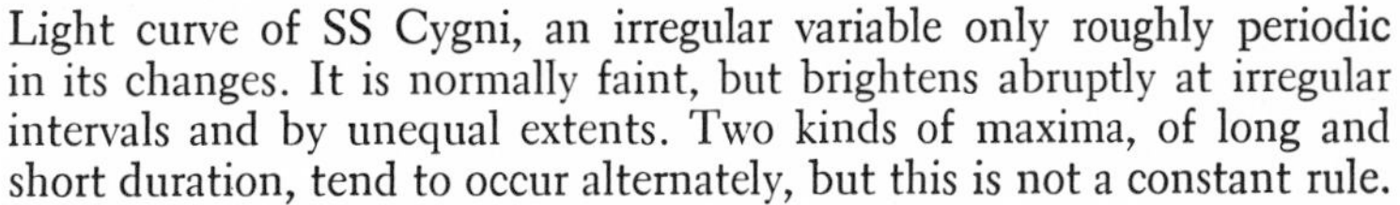
Light curve of SS Cygni, an irregular variable only roughly periodic in its changes. It is normally faint, but brightens abruptly at irregular intervals and by unequal extents. Two kinds of maxima, of long and short duration, tend to occur alternately, but this is not a constant rule.

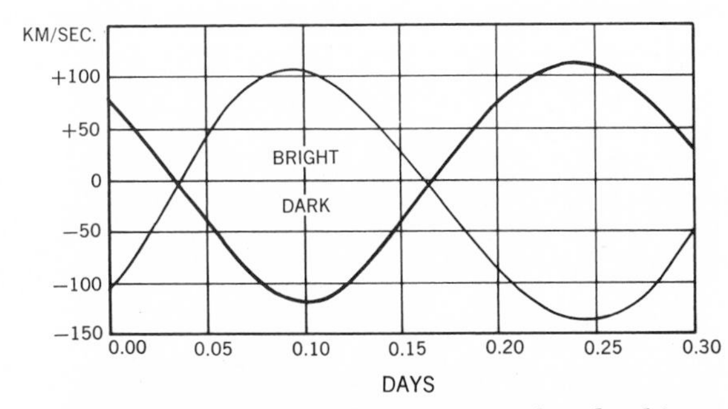

A. H. Joy's radial velocity curves for the binary star SS Cygni. When the dark-line star approaches us at its maximum speed, 115 kilometers per second, the bright-line star is receding at 122 kilometers per second.

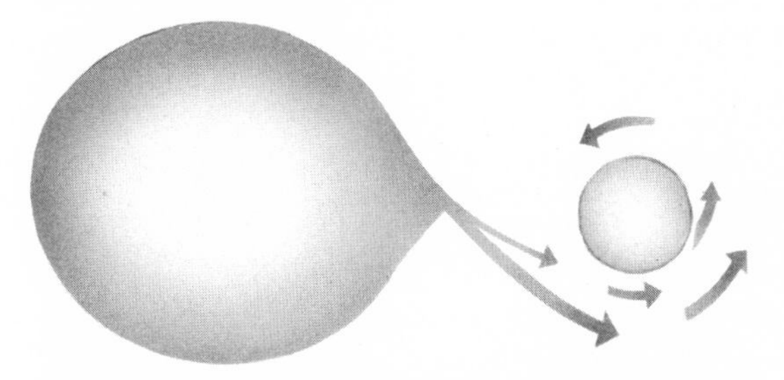

Model of a nova proposed by J. A. Crawford and R. P. Kraft. The giant red primary ejects gas into the region surrounding the hot white dwarf companion. The red star loses mass to the companion, which acts as the nova.

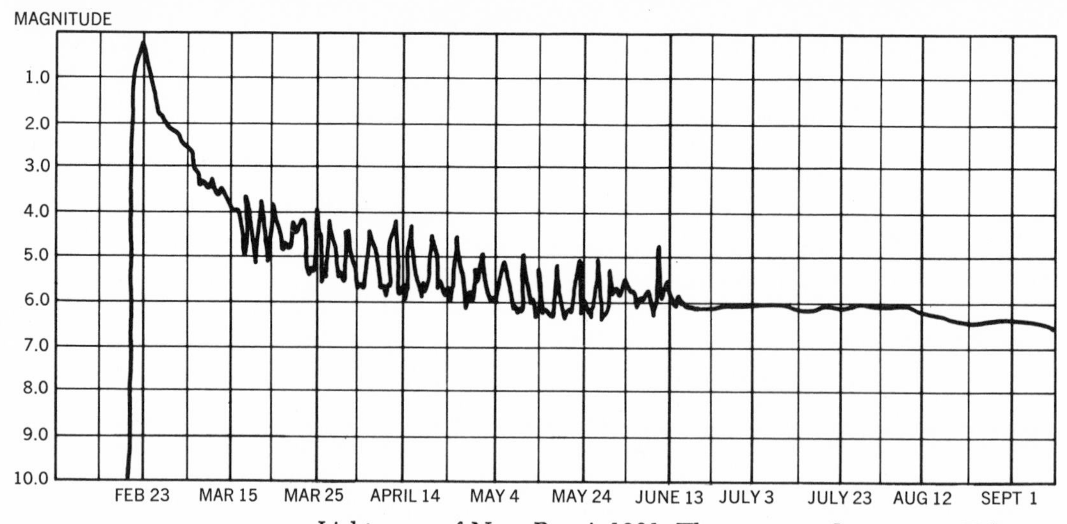

Light curve of Nova Persei, 1901. The nova was first seen on February 21 by T. D. Anderson, a Scottish clergyman. It attained its maximum brilliance (magnitude 0) on February 23, when it was about 160,000 times its prenova brightness of magnitude 13. Two weeks later a series of undulations set in and lasted for several months. The star continued to fade and did not reach magnitude 13 again until eleven years later.

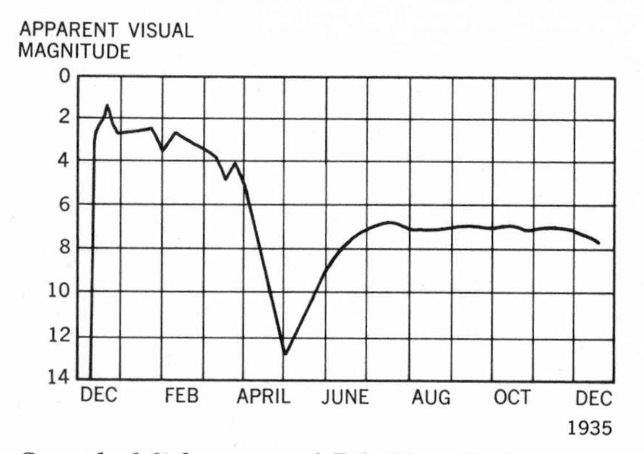

Smoothed light curve of DQ Herculis from December 1934 to December 1935. This nova was first seen on December 13, 1934, by J. P. M. Prentice, an English amateur astronomer. It attained its maximum brilliance (magnitude 1.3) on December 22, when it was about 300,000 times its prenova brightness of magnitude 15.

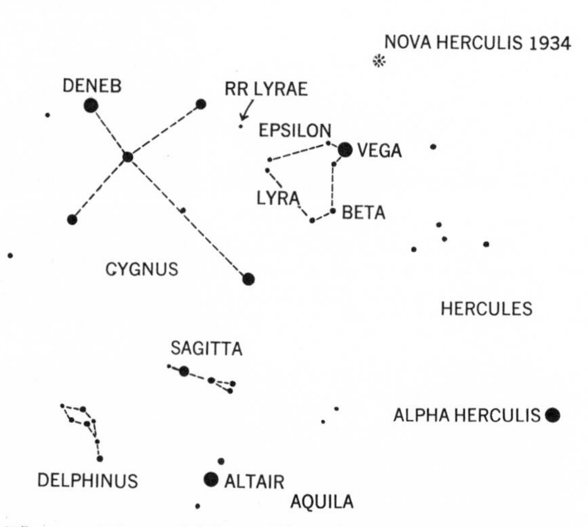

The position of Nova Herculis 1934 in relation to Vega and the Northern Cross (in Cygnus). Beta Lyrae, RR Lyrae, Epsilon Lyrae, and Alpha Herculis are also shown.

the final spectrum indicates that the star has become an extremely hot object surrounded by an expanding shell of gas. That this interpretation is substantially correct is shown by photographs of the nebulous envelopes around Nova Persei and Nova Aquilae. The envelopes grow larger with time, and at rates similar to those observed soon after the original explosions. Presumably the photosphere of a nova expands at such a great speed as to expel the overlying atmosphere. The photosphere then contracts, but the gaseous shell, having achieved escape velocity, continues its flight into space. If this is so, a normal nova outburst is only a "skin deep" affair. The star loses an almost negligible amount of its mass and is thus able to return to its prenova state.

In 1954 a series of photoelectric observations made at the Mount Wilson and Palomar Observatories led M. F. Walker to discover that Nova DQ Herculis was an eclipsing binary with a period of about 4.6 hours. The shape of its light curve is not unlike that of Algol-type variables and the eclipsing star is probably large, red, and low in luminosity. The nova itself appears to be a highly luminous white dwarf. It waxes and wanes in brightness with a period of only 71 seconds and is presumably pulsating. Since then Nova Persei 1901 and Nova Aurigae 1891 have also turned out to be binaries, while Nova Aquilae 1918 is suspect since its spectrum is reported to show radial velocity displacements. Whether all novae are binary systems remains to be seen, but the possibility that dwarf novae, recurrent novae and "ordinary" novae may have this particular feature in common is an intriguing one.

WOLF-RAYET STARS Another class of "exploding" star that claims part if not full binary-star status is the class of Wolf-Rayet stars. The first three members were discovered in 1867 by C. Wolf and G. Rayet of the Paris Observatory, and the number has now risen to well over two hundred. These stars have O-type spectra with strong radiation in the ultraviolet and are therefore extremely hot and luminous. Their spectra also have emission features in the form of broad, bright lines due to ionized helium, carbon, nitrogen, and oxygen. The lines are broadened by amounts which show that the

Expanding nebulosity around Nova Persei 1901 (200-inch Hale telescope). The star disks are not real but spurious, and those that appear to be in the nebulosity are either foreground or background objects.
—MOUNT WILSON AND PALOMAR OBSERVATORIES

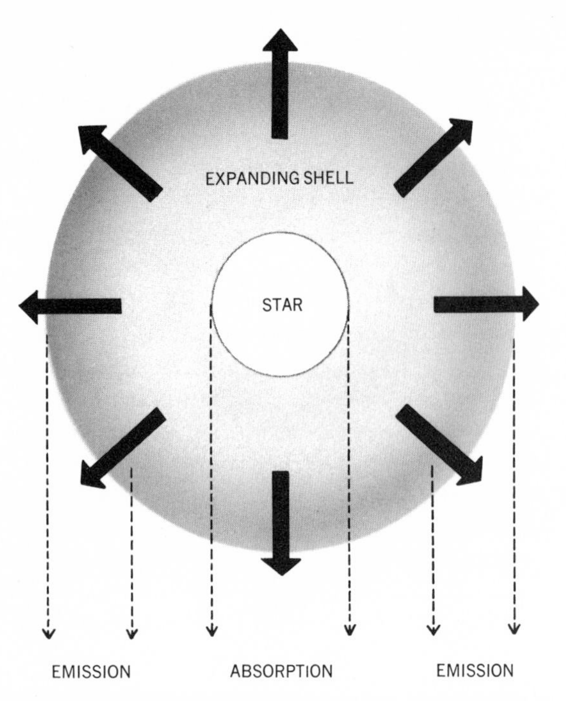

An expanding shell of gas surrounds a hot Wolf-Rayet star and accounts for the star's characteristic spectral features. In the direction of the observer the star's radiation undergoes selective absorption as it passes through the gas shell. The spectrum therefore contains dark lines, shifted toward the violet because this part of the shell is approaching the observer rapidly. The gas shell itself gives rise to bright emission lines, broadened owing to components of motion of the shell toward and away from the observer.

stars concerned are each surrounded by an expanding shell or shells of gas. A surprisingly large number of Wolf-Rayet stars have been found to be close spectroscopic binaries. In each of these the more massive and larger component appears to be a highly luminous dark-line O-star. The companion, the star with the gaseous shell, is even brighter. It is thought to have a surface temperature of the remarkably high order of 50,000 degrees centigrade and is certainly among the hottest stars known.

SUPERNOVAE Novae outbursts are quite mild compared with those of supernovae. In the latter, as we have already mentioned, the explosion affects an entire star, so no recurrence is possible. The most famous is the one of 1054, for its remains, the Crab Nebula, despite its distance of about 3,500 light-years, permit fairly detailed study by a variety of techniques. What was once a star is now an irregular mass of expanding gas, some 6 light-years across. The nebula is an extremely powerful source of X rays and radio waves, and it has been suggested that the former come from a thermal source in the form of a hypothetical object known as a neutron star. This is supposed to be a superdense stellar remnant whose core is up to 100 million times denser than that of a white dwarf. The star would be only a few miles across, would have a core temperature as high as 6 billion degrees, and would emit most of its radiation in the form of X rays.

Most astronomers seem to favor the view, first put forward by the Russian astronomer I. S. Shklovsky, that most of the radiation from the Crab Nebula is emitted largely by the synchrotron process and is therefore nonthermal. All attempts to detect an intensely hot central object have been unsuccessful. A fifteenth-magnitude star near the center has been investigated from time to time, but Kraft found that it had the spectral characteristics of a normal F-type star. According to him and others it is probably a background star quite unrelated to the nebula. Shklovsky, however, considers that the F star may be the remaining component of a binary system whose other component exploded to form the surrounding nebula.

There can be little doubt that naked-eye supernovae have appeared throughout human history. Of the bright objects seen in the sky and recorded by early Chinese astronomers, those of A.D. 185, 369, 827, and 1006 were probably supernovae. The outbursts of 1054, 1572, and 1604 (Kepler's star) were events in our own galaxy, but all the others since then have been extragalactic. S Andromedae 1885 occurred in the Great Galaxy in Andromeda, but because of its immense distance it remained a telescopic object even at maximum. Many of the more recent ones were discovered as a result of a systematic search maintained for many years by F. Zwicky of the Mount Wilson and Palomar Observatories. Zwicky's latest list, issued in 1964, contained a total of 152 objects.

Comparisons of the maximum absolute magnitudes of supernovae show that some are

10 to 100 times more luminous than others. The brighter, known as Type I, include the supernova of 1054 and reach an absolute magnitude of about −13. They are believed to develop from relatively old stars similar in mass to the sun. The fainter, Type II, reach an absolute magnitude of about −10 and develop from massive young stars.

HOT SUBDWARFS The hot, white stars of dwarf novae and ordinary novae form a class known as the "hot subdwarfs." They lie on the left-hand side of the H-R diagram, above the white dwarfs and to the left of the main sequence. They may represent the predwarf phase in stellar evolution for their store of central hydrogen is depleted and they maintain their energy output by converting helium into carbon. But until the cause of their outbursts is known we cannot tell whether these are related to their binary nature. Novae certainly do not become binaries as a result of the outbursts. The latter are too superficial in nature to bring about the fission of one star into two. Presumably the more massive component at first sheds material to its companion and also into space, so its evolution is speeded up while that of the companion is slowed down. Eventually the star that was originally more massive becomes a white dwarf, by which time the companion has evolved into a red giant. The latter then loses material to the white dwarf and thus gives rise to the dwarf novae we see today.

Origin of Binaries

If the great majority of stars were single objects it would be reasonable to assume that they were all born as such. The high frequency of binaries has led some astronomers to question the validity of this assumption, but it may still be substantially correct if binary systems are formed by the division of a single star into two components. Some forty-five years ago the English cosmologist J. J. Jeans suggested that the division or fission was brought about by excessive rotation. A star, he suggested, shrinks as it ages, and by the principle of the conservation of angular momentum, spins faster and faster. His mathematical investigations showed that if a star behaves as a liquid mass, excessive speed of rotation causes it to become highly elongated or cigar-shaped. At still higher speeds of rotation the mass becomes pear-shaped. A waist or neck forms and deepens until it cuts the body into two separate parts. The parts then revolve about one another in an atmosphere of ejected matter and constitute the components of a double star. If the components themselves shrink with increasing age, and to the extent of producing further fissions, the system becomes a multiple one. But for most close pairs further fissions would be delayed by the tidal action of one star on the other. This would tend to equalize their periods of rotation and revolution, and their orbits would become larger and more elliptical.

Rapidly Rotating Stars

Since Jeans' time spectroscopic studies have revealed that certain O- and B-type stars have high speeds of rotation. The sun's equatorial speed is just over a mile a second, but these stars turn at speeds of up to about 340 miles a second. The speeds tend to increase with mass and are often fast enough to spin surface material into space. The spectra of rapidly rotating stars show bright-line emission features due to extended spirals or rings of hot gaseous material that are being whirled away from their equatorial regions.

The stars concerned are known as Of and Be stars, the letters f and e referring to the emission peculiarities of their spectra. A third group with similar characteristics is named after the prototype P Cygni, a nova which flared up in 1600. Its spectrum shows strong emission lines with violet absorption borders, thereby showing that at least a part of its atmosphere is expanding.

The high velocities of spin recorded so far are well below those required for fission. Yet the stars concerned are massive objects on or near the main sequence, so they are more likely to be expanding than contracting. In the light of modern ideas of stellar evolution, contraction is important only in the early pre-main-sequence stage before thermonuclear reactions set in. Binaries whose components are so similar as almost to be twins are found all along the main sequence. Their components are presumably similar in age, have gone through a similar development, and have been together since birth. But how their association began, whether from the parent gas cloud as two condensing spheres of gas or by the fission of a newly born star, remains a mystery.

Evolution of Components of Binaries

Binaries whose components are moderately to widely spaced are usually of spectral types G, K, and M and are much less frequent than their closer brethren. They probably represent the components of what were once close twinned pairs, for a pair of orbiting stars is bound to separate gradually if only through the attractions of other stars. At the same time the interchange of material between them could profoundly affect their individual rates of evolution and eventually make them widely dissimilar. In cases where the original masses of close components are considerably different, the primary, in evolving toward the giant stage, might engulf its smaller companion, thereby leaving just a single star where once there were two. J. A. Hynek of Dearborn Observatory has suggested that if this happened it would change the subsequent evolution of the primary to such an extent that it would never become a normal giant. This might account for the observed absence of short-period spectroscopic binaries among giant stars.

CHAPTER VIII

PULSATING STARS

When Jeans looked for observational support for the idea of the fission of one star into two he thought that he found it among the Cepheids. These highly important variable stars are named after Delta Cephei, a naked-eye star whose light fluctuations were first detected and studied by Goodricke in 1784. The general shape of their light curves, the suspected decreases in the lengths of their periods, and changes in their spectra all lent color to the notion that they were young stars in the process of fission. About 1915, however, H. C. Plummer and Shapley independently suggested that the Cepheids were single pulsating stars, and a little later Eddington put the idea to the test of mathematical analysis.

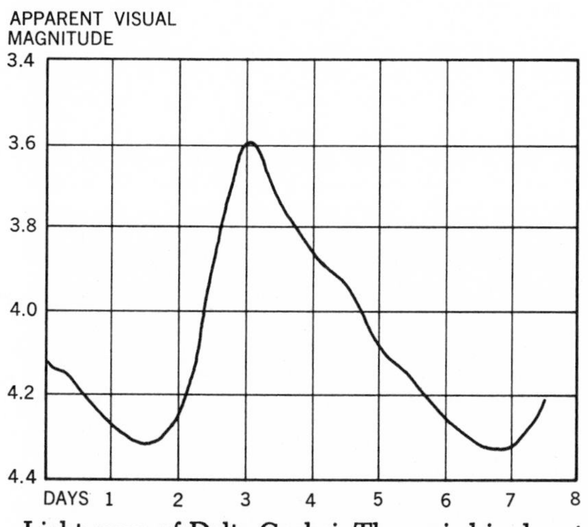

Light curve of Delta Cephei. The period is almost constant over long intervals of time, the rise to maximum light fairly steep, and the decline to minimum relatively short.

Cepheids

The light curves of Cepheids have characteristics quite different from those of eclipsing variables. The change in brightness is continuous, usually with a fairly steep rise to maximum and a more gradual descent to minimum, and the period can be anything from less than a day to about 100 days.

Delta Cephei itself varies between magnitudes 3.6 and 4.3 in a period of 5.37 days. The rise to maximum is accomplished in about a third of this time and is quite smooth. The fall to minimum, however, is accompanied by small irregularities. The light curve of Eta Aquilae has a similar shape, but there is a distinct hump or shoulder about midway between maximum and minimum. This Cepheid, the second to be discovered (by N. Pigott in 1784), has a period of 7.18 days and varies between magnitudes 4.0 and 5.0. Many others have similar shoulders on their light curves, while some, like Zeta Geminorum (period 10.15 days) have almost symmetrical light curves.

CHARACTERISTICS OF VARIABILITY When the light curves of these and other Cepheids are arranged in order of increasing period, they show a definite trend in shape. Those of short period have steep narrow maxima, but with longer periods the curves tend to become broader. They then develop a shoulder on their declining sides and finally appear fairly smooth and symmetrical at the longest periods. This, however, is only a general tendency or trend, not a fixed relationship, and two Cepheids of the same or similar periods can have light curves of different shapes.

Quite a number of Cepheids remain constant in their light periods for many years and then change abruptly to take up another steady rhythm. A striking example is RZ Cephei. Its period of just over 7 hours 24 minutes became suddenly shorter by 3.98 seconds in 1901 and then abruptly increased by 4.33 seconds and 1.84 seconds in 1916 and 1923 respectively. Other well-observed examples are Eta Aquilae, Zeta Geminorum, and Polaris. The last, the brightest Cepheid in the sky, had a sudden increase in period of 2.3 minutes around the year 1928 but has remained steady ever since. Actually this star is a triple system with the brightest component a Cepheid of light period just under 4 days.

In the spectrum of a Cepheid the dark lines are periodically shifted slightly toward the red and violet ends of the spectrum. The effect is similar to that observed with spectroscopic binaries in which one component completely outshines the other, thereby providing only one set of observable spectrum lines. The shifts are clearly due to velocities of recession and approach, but the shape of the light curve does not fit the picture of a pair of mutually eclipsing stars. Maximum line-shift toward the violet, indicating maximum speed of approach, occurs near maximum brightness. But maximum line-shift toward the red, or maximum speed of recession, occurs near minimum brightness; if the star were an eclipsing binary the maximum shift would coincide with maximum brightness. The changes therefore permit only one possible explanation—they are intrinsic and arise from the periodic swelling and shrinking of a single star.

SPECTRAL CHANGES Contrary to expectation, a Cepheid does not reach maximum brightness when its size is greatest but when it is expanding most rapidly. Similarly, minimum brightness occurs when it is contracting most rapidly. But much more than a comparatively small change in size is involved. As a Cepheid increases in brightness, changes in its color index and in the relative strengths of various lines in its spectrum indicate that the surface temperature is rising. In some instances the increase is so great that the star changes spectral type by nearly one whole class of the spectral sequence. The spectrum is not precisely similar to that of a normal nonvariable star. At maximum light it generally resembles that of supergiants of spectral types F5 to F8, while at mini-

The Small Cloud of Magellan. Two galactic globular clusters are in the foreground. The one at the top right corner is 47 Tucanae.
—HARVARD OBSERVATORY

mum it generally resembles that of supergiants of types F8 to K1. Nearly all the change in brightness can therefore be laid at the door of a rise and fall in surface temperature, the atmosphere of the star often being about 1,000 degrees centigrade hotter at maximum light than at minimum.

For the majority of Cepheids the range in spectral type, and therefore in temperature and brightness, increases with increasing period. Polaris, for instance, stays at spectral type F7 and varies between the narrow magnitude limits of 2.08 and 2.17. SV Vulpeculae, with a much longer period of 45.13 days, changes from type G2 to K5 as its brightness decreases from magnitude 8.43 to 9.40. Cepheids with short periods tend to be hotter than those with long periods. This tendency, inferred from a corresponding trend in spectral type, is known as the period-spectrum relation.

PERIOD-LUMINOSITY RELATION Cepheids are invaluable as distance indicators. Their function in this field is due to a relation between their periods and luminosities. That the two attributes are related first came to light in 1912. Miss H. S. Leavitt of the Harvard Observatory discovered twenty-five Cepheids in the Small Cloud of Magellan. She arranged them in order of increasing period and was surprised to find that they formed an order of increasing brightness. Even more important, she succeeded in establishing a general relationship between period and average magnitude. Since the stars concerned were all at roughly the same distance, the relationship was also one between period and absolute magnitude, or alternatively, period and luminosity.

In Shapley's time, some fifty years ago, the Clouds of Magellan were thought to be very remote systems of stars, but their distances were unknown. So before a period-luminosity relation could be established it was necessary either to determine these distances or to extend the sample so that it could include Cepheids at known distances. Unfortunately the Cepheids in our galaxy are too far away to permit accurate measurement of their trigonometric parallaxes, but by using proper motions and other indirect data, Shapley eventually arrived at a fairly reliable period-luminosity relationship. He then used it to determine the absolute magnitudes and distances of Cepheids in globular clusters, and was able thereby to outline the tremendous extent of the Milky Way System.

Other Pulsating Stars

Globular clusters are rich sources of variable stars whose periods range from about an hour and a half to little over a day. Their light curves are Cepheidlike in character but many have extremely rapid rises to maximum and relatively prolonged minima. They were first detected in

Messier 13 in Hercules, the brightest globular star cluster in the northern sky. It looks like a hazy star to the unaided eye and was discovered accidentally by Halley in 1714.

—MOUNT WILSON AND PALOMAR OBSERVATORIES

1895 by S. I. Bailey on photographs of certain globular clusters. Many more were found in other globular clusters, so for a time they were called "cluster-type variables." At the beginning of the present century, however, Mrs. W. P. Fleming of the Harvard Observatory discovered that RR Lyrae, outside a globular cluster, also behaved like a cluster variable. This star, the brightest of its type (apparent magnitude 6.8 at maximum) has a period of 13.6 hours and a light-range of 1.1 magnitudes. So many other "outsiders" have been found since that cluster variables are now generally known as "RR Lyrae stars."

RR LYRAE STARS RR Lyrae stars, like Cepheids, are pulsating stars. They also change in spectral type, so their light fluctuations are almost entirely due to changes in surface temperature. Their spectra are more abnormal than those of Cepheids. Struve has remarked that at minimum light the hydrogen lines in the spectrum of RR Lyrae indicate a spectral type F6 whereas the metallic lines indicate type F0. At maximum light the hydrogen lines indicate type F0 and the metallic lines type A2. Similar peculiarities are found among other members of the class, so they do not fit into the framework of the normal spectral sequence. Unlike the Cepheids they show no apparent period-spectrum or period-luminosity relation. On the contrary, all RR Lyrae stars are of (average) absolute magnitude 0.5. This remarkable property makes them about one hundred times more luminous than the sun although generally much fainter than Cepheids. They can therefore be seen over great distances and are a powerful tool for surveying our own and nearby galaxies.

The light curve of RR Lyrae has a curious property. Its form alters progressively from one cycle to the next and goes through one complete change in 72 cycles. In addition, the length of the period undergoes a slow oscillation. A detailed analysis of this change shows that the star's principal period of light variation is accompanied by several harmonics. The situation is analogous to that of a violin string, which when bowed in a certain way, produces its fundamental note along with several overtones.

Other RR Lyrae stars behave in a similar way, as also does a small but growing number of hot A- and F-type stars sometimes called "dwarf Cepheids." The latter include very short-period variables like SX Phoenicis (1.3 hours) and Delta Scuti (4.7 hours). They are dwarfs in the sense that they lie on or near the main sequence and are therefore much less luminous than the giant and supergiant Cepheids.

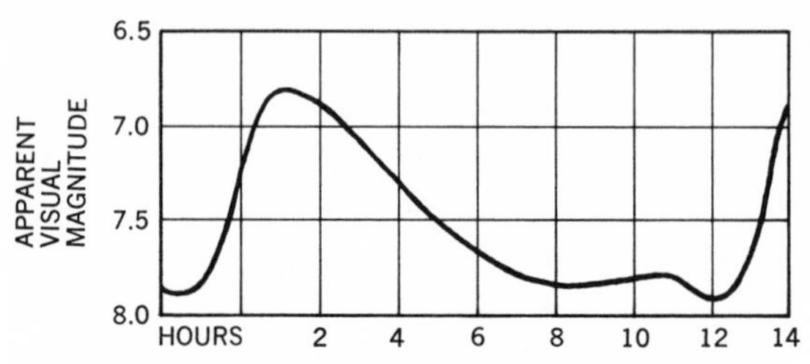

Sample light curve of RR Lyrae, a pulsating star which brightens rapidly and fades slowly in a fundamental period of 13.6 hours.

BETA CANIS MAJORIS VARIABLE STARS Short period and high surface temperature are also characteristic of a distinct class of pulsating star named after Beta Canis Majoris, its brightest member. These stars, all hot, massive B-type giants, have small but regular variations in brightness. Their periods range from 3½ hours (Gamma Pegasi) to 6 hours (Beta Canis Majoris) and are fairly closely associated with their spectral types. Those of spectral type B0 are more luminous and have longer periods than those of spectral type B3. Several of them have been found to have increasing periods. The period of BW Vulpeculae, for example, is increasing steadily by about three seconds a century, a relatively large rate. The rates of some of the others are only 0.1 to 0.2 second a century, but these values, although small, have a sound observational basis.

Struve thought that the observed increases in period might represent evolutionary changes. He suggested that once a hot, massive star left the main sequence it expanded and became a Beta Canis Majoris variable. Theory then indicated that the increase in volume and consequent decrease in average density would gradually slow down the period of light variation. As he pointed out at the time (1958), if the changes were of this type they would provide astronomers with an extremely sensitive method of detecting and following evolutionary changes in individual stars.

W VIRGINIS It would be unusual, to say the least, if all Cepheids had characteristics like

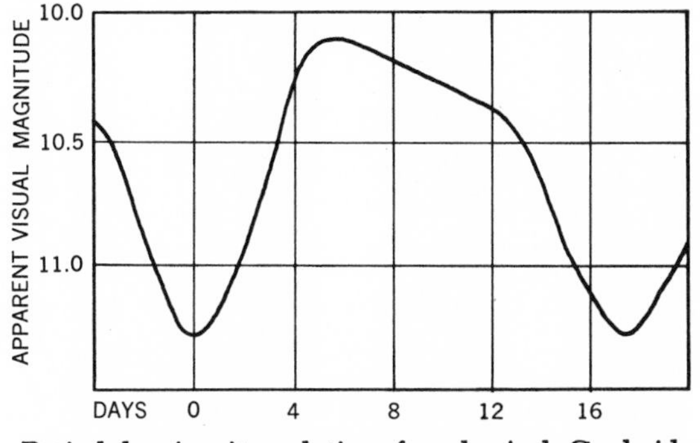

Period luminosity relation for classical Cepheids (Population I) and Cepheids of the W Virginis type (Population II).

those of Delta Cephei. A large number are definitely nonconformists, and to such an extent as to form at least one breakaway class. The most conspicuous among them is W Virginis, a ninth-magnitude yellow star whose period (17.27 days) and fairly steady light curve gives it Cepheidlike qualities. During its rise to maximum light, bright lines due to hydrogen appear in its spectrum and for a short time two absorption spectra are seen simultaneously. It is undoubtedly a pulsating star, but in addition it is affected by rhythmic surges which break through and away from its surface. Similar spectral peculiarities are shown by some RR Lyrae stars, but never by the so-called "classical Cepheids."

POPULATION II CEPHEIDS W Virginis is a typical specimen of a class of pulsating variable known as "Population II Cepheids." These stars are found in globular clusters, nearby galaxies, and our own galaxy, but in the last case they tend to lie well clear of the Milky Way. This is also true of classical Cepheids, but with one important difference: those belonging to our galaxy lie in and near the Milky Way. In other words, the two classes of Cepheid are distributed differently in relation to the plane of the Milky Way and appear to represent two different stellar populations. Population II Cepheids are all about 1.5 magnitudes less luminous than the classical, or Population I, Cepheids. They have their own period-luminosity law and those with periods of a day or so show every indication of being closely related to, if not identical with, RR Lyrae stars.

RV TAURI It is now fairly well established that Population II Cepheids link up with a group of orange and orange-red giant stars named after RV Tauri, its brightest member. These stars have periods which range from about 40 to 150 days, but those near the upper end of the range tend to be irregular in their light behavior. In general, however, they subscribe to a period-spectrum and a period-luminosity relation. They have spectral characteristics similar to those of RR Lyrae and W Virginis stars, occur in globular clusters but are not confined solely to them, and are found in parts of the sky far from the region of the Milky Way. They extend the sequence of RR Lyrae stars and Population II Cepheids to the region (on the H-R diagram) of the giant red variables.

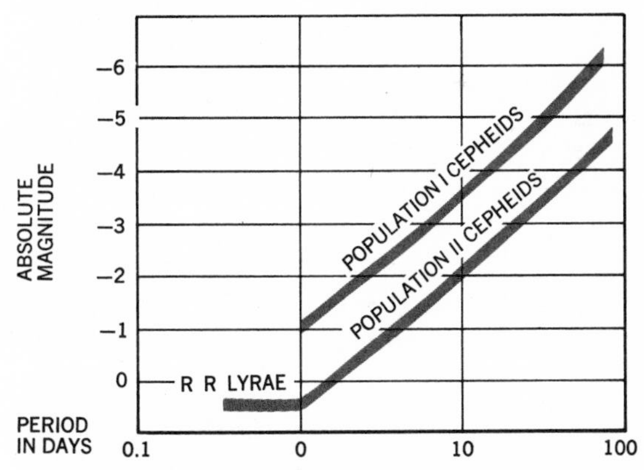

Light curve of W Virginis, a star with a steadiness of light variation similar to that of long-period Cepheids, but less smooth in its decline from maximum brightness.

Semiregular and Long-period Variable Stars

The giant red variables form a most interesting and important class. Some of them vary in brightness rather irregularly in cycles of 40 to

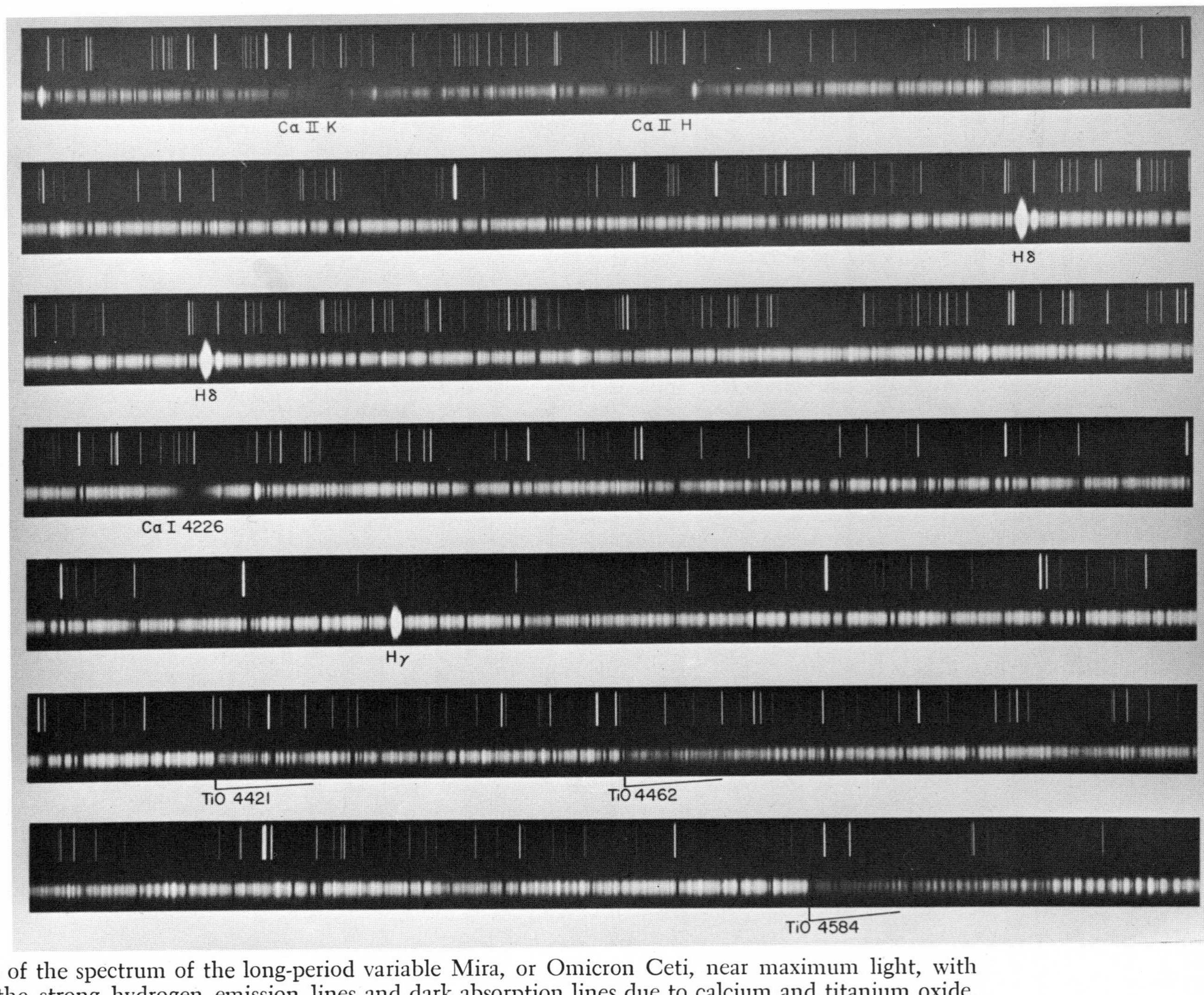

Representative portions of the spectrum of the long-period variable Mira, or Omicron Ceti, near maximum light, with high dispersion. Note the strong hydrogen emission lines and dark absorption lines due to calcium and titanium oxide.
—Lick Observatory

100 days and are called semiregular variables. Betelgeuse and Antares come in this group, but their periods, or rather, cycles, are of the order of several hundred days. Two others are Eta Geminorum and Mu Cephei, Herschel's "garnet" star. The great majority of red variables fluctuate more rhythmically and with greater amplitudes in periods that range from about 100 to 1,000 days. They are known as long-period variables and consist mainly of M-type giants of low average density, although some are of spectral types N, R, and S. Their prototype is Omicron Ceti, discovered by D. Fabricius in 1596 and named by him "Mira," the "wonderful." This star varies in brightness from the third to the ninth magnitude in a period of about 331 days, but these are average values —the ranges in magnitude and in the period are both subject to large variations.

The maximum intensity of the radiation from these red giants and supergiants lies in the far red or near infrared. An increase in temperature of only a few hundred degrees is sufficient to shift the maximum intensity from the infrared into the visible red part of the spectrum. As a result the star brightens considerably, although the increase in its total radiation is comparatively small. Chi Cygni, for example, has an average brightness range of eight magnitudes, but the bolometric range is only about one magnitude. At its minimum, Chi Cygni is roughly of the thirteenth magnitude, so the large increase brings it into naked-eye visibility, a feature which insured its early discovery, by Kirch in 1686. Many other red variables, quite faint at and near their minima, have a similar range in magnitude. So when they are observed with a small telescope, or even a pair of binoculars, their changes, although slow, can be quite striking.

SPECTRUM CHANGES IN GIANT RED VARIABLES The increase in the brightness of a long-period variable is due largely to the increase in its surface temperature. At minimum the dark bands in its spectrum due to titanium oxide (type M), zirconium oxide (type S), and compounds of carbon (types R and N) are strong, indicating low temperature. The heavy bands effectively reduce the overall brightness of the spectrum and thus one can conclude that the compounds they represent reduce the luminosity of the star itself. As the temperature increases the absorption bands weaken and bright lines of hydrogen and some of the metals appear, to reach their maximum intensity at or soon after maximum brightness. The bright lines then gradually fade and disappear before minimum is reached. Yet although they vary in prominence with the light changes, they can sometimes lag behind by almost a sixth of a period.

These stars are thought to have veil-like clouds of metallic vapors in their distended atmospheres. The clouds thicken and increase their coverage as the star grows cooler, but as the temperature rises they tend to disperse, exposing the upper part of a hot hydrogen zone beneath the normal photosphere. Theoretical studies by Schwarzschild (1937) indicate that only the interior of the star pulsates as one whole. The pulsations set up waves of compression and rarefaction in the outer layers, whose state of pulsation therefore lags behind that at much deeper levels. This might account for the difference between the time of maximum overall brightness and the time when the hydrogen lines reach their maximum brightness.

Photographs of the infrared parts of the spectra of several red giants have recently been taken with the balloon-borne Stratoscope II telescope. They show that Mira has a large amount of water vapor in its atmosphere. Betelgeuse has less than one tenth as much as Mira, while the stars Mu Cephei, R Leonis, Rho Persei, and Mu Geminorum have quantities intermediate between the two extremes. The vapor will absorb substantial amounts of the radiation from the photospheres of these stars and will therefore affect the temperatures of their atmospheres.

Giant red variables, like several other classes of stars, have their wayward members. It might even be said that they are all more or less wayward, for their light behavior and spectral peculiarities are both complex and puzzling. They are generally considered to be pulsating stars, but the changes in radial velocity are small in range and difficult to measure. Many have quite irregular light variations and are therefore unpredictable in their changes. Some stay fairly regular for many years and then suddenly brighten in novalike outbursts. When these occur the usual M-type spectrum is accompanied by bright emission lines similar to those seen in gaseous nebulae and requiring high excitation energies.

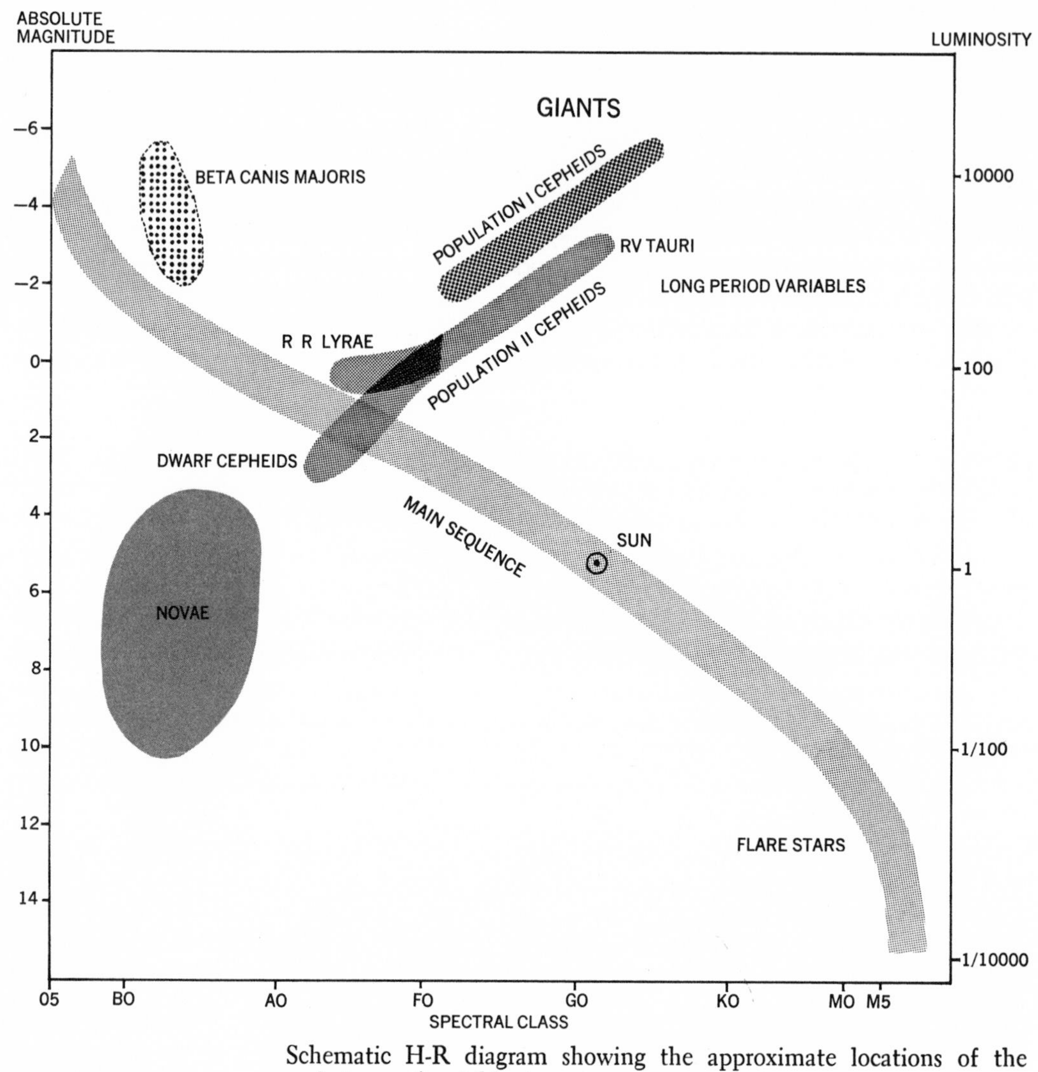

Schematic H-R diagram showing the approximate locations of the main types of variable stars.

Symbiotic Stars

Stars whose spectra show this extraordinary combination of low- and high-temperature features are called symbiotic stars. They have been extensively studied and discussed by P. W. Merrill of the Mount Wilson and Palomar Observatories and form an important group of about twenty-five known stars. The prototype is Z Andromedae, a faint M-type star of about the eleventh magnitude which suddenly brightened in the years 1901, 1914, and 1940. After each outburst the brightness gradually declined, but in a series of rhythmic fluctuations similar to those shown by declining novae. Other symbiotic stars are R Aquarii, AX Persei, CI Cygni, BF Cygni, and AG Pegasi. In the case of R Aquarii, photographs taken with the 100-inch telescope on Mount Wilson have revealed that the star is embedded in or associated with irregular nebulosity of considerable extent.

The most obvious explanation of the spectral features is that a symbiotic star is a binary system; both components—one an M-type giant, the other a hot subdwarf—are involved in streams of gas. But there is, as yet, no supporting observational evidence. The light variations and spectral features are so complex that they obscure effects due to mutual eclipses and orbital velocities. Merrill suggests that the spectral peculiarities in AG Pegasi might be produced by a revolving jet of gas pulled around by one component of a binary system confined within an expanding gaseous envelope. Another view is that the peculiarities arise from a single pulsating star disturbed by enormous flare and prominence activity and possessing a very bright and hot corona.

ETA GEMINORUM Can a star be at once a visual binary, semiregular variable, and an eclipsing binary? Strangely enough, Eta Geminorum fulfills all three conditions. As a visual binary it consists of an M3 giant primary and G8 giant secondary. The primary is a well-known semiregular variable, but spectroscopic observations indicate that it is also an eclipsing binary with a period of revolution of about eight years. The components are thought to be two giant M stars extremely close together. Finally, recent spectroscopic observations by A. J. Deutsch have shown that all three stars are enveloped in an enormous expanding cloud of cool gas. Here, then, is further evidence of the gradual loss of mass of a red giant by means of an expanding envelope. If the process is common to all red giants and supergiants it must clearly play an important role in stellar evolution. For, as we saw earlier, a major requirement in the modern picture of stellar evolution is that many stars lose a considerable part of their original mass somewhere between the main-sequence and white-dwarf stages.

CHAPTER IX

STARS AND NEBULAE

During the course of his surveys Herschel came across a number of faint objects which he called "cloudy or nebulous stars." He thought at first that they were star clusters so remote as to be beyond the resolving powers of his telescopes. But as his experience widened he began to suspect that this interpretation was incorrect. The "cloudy stars," small, compact, and roundish, looked different from collections of faint stars. One of them, picked up toward the end of 1790, consisted of a star "of about the 8th magnitude" at the center of "a faint luminous atmosphere of a circular form and of about 3′ of arc in diameter." Since the "atmosphere" obviously belonged to the central star it could not be composed of stars. There was only one possible conclusion. The object in question consisted of a star immersed in "true nebulosity" or "shining fluid, of a nature totally unknown to us."

Some of Herschel's cloudy stars have since turned out to be remote galaxies with bright nuclei. Others are stars associated with clouds of gas and dust located in the general plane of the galaxy. Others again are known as planetary nebulae: they generally present faint greenish disks in telescopes of moderate power and bear a close resemblance to the planet Uranus. The object described by Herschel was a planetary nebula in Taurus. It owed its appearance to the fact that its central star, of magnitude 9.7, is unusually bright in relation to the surrounding nebulosity.

The Nature of Nebulous Material

Herschel suggested that the "shining fluid" represented material in a primitive state which would ultimately condense into stars and planets. Others, including his son, John F. W. Herschel, and Lord Rosse, thought otherwise. They examined a number of patches of "true nebulosity," among them the Great Nebula in Orion, and thought that they could detect signs of resolution into stars. The answer to the problem came in 1864, when W. Huggins attached a spectroscope to his telescope and directed it to a planetary nebula in the constellation of Draco. His account is worth repeating.

> *I looked into the spectroscope. No spectrum as I expected! A single bright line only! At first I suspected some displacement of the prism and that I was looking at a reflection of the illuminated slit from one of its faces. This thought was scarcely more than momentary; then the true interpretation flashed upon me. The light of the nebula was monochromatic . . . a little closer looking showed two other bright lines towards the blue. . . . The riddle of the nebulae was solved. The answer which had come to us in the light, itself, read: not an aggregation of stars, but a luminous gas.*

Huggins tried to identify the three lines but without success. The one in the green was close to a prominent line in the spectrum of glowing nitrogen, but the other nitrogen lines were absent. He argued that the nebular gas could not be the primal stuff of stars, otherwise it would give rise to sets of lines due to a large number of elements. The three bright lines also appeared in spectra of the Great Nebula in Orion and other nebulae. But as they completely defied identification they were ascribed to an undiscovered element appropriately called nebulium. Later studies showed that the light from gaseous nebulae is produced under conditions that cannot be duplicated in a terrestrial laboratory. These conditions, like those in the sun's corona, are characterized by extremely low densities and very high levels of excitation. As a result most of the elements present are ionized, and the prominence of their spectral lines cannot be taken as an index of their relative abundances.

In 1928 I. S. Bowen showed that the so-called nebulium lines were due mainly to doubly ionized atoms of oxygen. Additional lines are produced by neutral hydrogen and also by hydrogen, oxygen, nitrogen, neon, carbon, and sulfur in several stages of ionization. Yet here, as elsewhere in the universe, hydrogen is easily the most abundant element; the other elements exist only as "impurities."

Herschel's planetary nebula is NGC 1514, and the one observed by Huggins is NGC 6543. These designations refer to places in the *New General Catalogue*, a list of the positions of 7,840 nebulae and star clusters prepared by J. L. Dreyer of Armagh Observatory and published in 1888. The catalog was later supplemented by two IC, or *Index Catalogues*, of a further 5,086 objects, but those already included in Messier's earlier list are still referred to by their Messier numbers. Of these thirty-four are galaxies, fifty-seven are star clusters, seven are diffuse or irregular nebulae, and four are planetary nebulae.

Planetary Nebulae

More than five hundred planetary nebulae are known at present. Many have a definite planetary appearance, being circular and fairly uniform in surface brightness. Some, like M57, the well-known "Ring Nebula" in Lyra, look like elliptical smoke-rings. Others, composed of loops and knots, have complex structures and peculiar shapes. Yet practically every one has a blue star at the center of whatever symmetry there may be. This very hot star is the activating agent, so to speak. Its radiation, most intense in the ultraviolet, ionizes the atoms in the gas. The resulting free electrons then recombine with ions (positively-charged atoms) and the gas becomes luminous, emitting light at certain specific wavelengths. This process, in which radiation of short wavelength is absorbed and re-emitted at longer wavelengths, is known as fluorescence or luminescence.

Typically, the visible part of a planetary nebula takes the form of a shell or bubble with a hot O-type star at its center. The smoke-ring appearance is no more than an effect of perspective. Many of the central stars have Wolf-Rayet characteristics, and this suggests that planetary nebulae are stars which have ejected, or continue to eject, clouds of gas. That the gas is moving is shown either by a doubling or a broadening of the spectral lines, but this effect on its own does not tell us whether the movement is away from the star or toward it. Fortunately, a comparison between modern photographs and others taken about fifty years ago has removed all doubt in the matter. Some planetaries have definitely increased in overall size, so their gaseous shells must be expanding.

Planetary nebulae are relatively distant objects. The nearest is NGC 7293, the ring nebula in Aquarius, at an estimated distance of roughly 300 light-years. It is the only one to show an appreciable trigonometric parallax. It is also the largest known planetary, with an apparent diameter of 15 minutes of arc, or about half that of the full moon; its actual diameter is about 1.3 light-years. The ring is double in

NGC 7293, the Ring Nebula in Aquarius, photographed in red light.
—MOUNT WILSON AND PALOMAR OBSERVATORIES

M97, the Owl Nebula in Ursa Major.
—MOUNT WILSON AND PALOMAR OBSERVATORIES

places so its material was probably ejected at different times or at different rates. Another planetary is a member of the globular cluster M15, whose distance has been established from its short-period variable stars. But apart from these two, planetaries lie at distances which are only roughly known, if known at all. This brings about a corresponding uncertainty in our knowledge of their real diameters and the absolute magnitudes of their central stars. Comparatively little is known about their nature, probable ages, and evolutionary significance.

The second largest planetary is M27—NGC 6853—popularly called the Dumbbell Nebula. It lies close to the star 14 Vulpeculae and in small telescopes looks like two separate nebulae connected by a luminous bridge. Photographs with large telescopes show that the two main masses merge into one another, giving the nebula a roughly oval shape. The third largest planetary is M97, or NGC 3587, a nearly circular but faint object in Ursa Major. It is fairly uniform in brightness except for two circular dark patches, first seen by Lord Rosse with his 72-inch reflector. Its overall appearance is reminiscent of the face of an owl, hence it is often referred to as the Owl Nebula.

THE RING NEBULA IN LYRA. The best-known of all planetaries is M57, or NGC 6720, the Ring Nebula in Lyra. It is probably the fourth in order of apparent size, but its broad elliptical ring, distinct and regular in shape, makes it a striking object in telescopes of moderate size. Photographs taken with large telescopes show that the ring has a distinct filamentary structure and completely outshines the faint but intensely hot central star. Those taken at particular wavelengths show that doubly ionized oxygen and neutral hydrogen are detectable in the shell, but that ionized helium and neon can be detected both in and within the shell. This is not surprising since helium, neon, and other inert gases are hard to ionize and so retain some of their electrons despite the very high energy levels of the radiation in the immediate vicinity of the exciting star.

ORIGIN OF GASEOUS SHELLS There is evidence that the central stars of planetary nebulae are hot subdwarfs. They certainly have high surface temperatures and probably outshine the sun several times. A typical central star would be about ten times smaller than the sun in diameter and have a high average density. If this is so the star, formerly a red giant, has exhausted most if not all of its nuclear fuel, undergone collapse of its core, and thrown off its outer layers in the form of a shell. The shell will continue to expand but will gradually lose its identity as its material spreads over an ever increasing volume of space. Meanwhile, the central star will continue to contract through the white-dwarf stage toward ultimate extinction.

The postnova interpretation of planetary nebulae must not be taken too far. What little information we have shows that the rate of expansion of a planetary shell is much smaller than that of a nova shell. The ejection of material in a planetary is a relatively slow process, and in this respect planetaries are similar to Wolf-Rayet stars. In at least one case, that of NGC 2392, the Eskimo Nebula, the central star is believed to be still ejecting material. Measurements of the spectral-line widths indicate fairly large radial velocities, but the nebula has undergone no perceptible change in shape or size in nearly fifty years.

If novae are responsible for planetaries,

why do we not find a planetary nebula on the site of Tycho's supernova of 1572? The nearest, NGC 7635, is about 8 degrees away!—though it looks unusual and could be the remains of a nova. Old chronicles record the appearances of two earlier novae in this part of the sky—one in 945 and another in 1264. A weak radio source occupies the estimated position of Tycho's supernova, but no definite remnants can be detected. Much the same can be said of the supernova of 1604 in Ophiuchus, whose site lies in a region heavily obscured by dark nebulosity.

THE CRAB NEBULA Lists of planetary nebulae usually include M1, or NGC 1952, the Crab Nebula in Taurus. In small telescopes this object looks like a misty patch of light, roughly oval in shape, with a brightness that increases toward the center. It looks much the same on ordinary photographs taken with large telescopes, but the boundary is then seen to be quite irregular. Photographs taken in the red light of hydrogen reveal that the outer part consists of a complex system of contorted filaments, most of which run radially from the center. The main peculiarity is the intense continuous spectrum given by the inner amorphous mass. This contains nearly all the energy emitted by the nebula and is best explained in terms of the synchrotron mechanism of high-speed electrons moving in a strong magnetic field. It is crossed by numerous bright emission

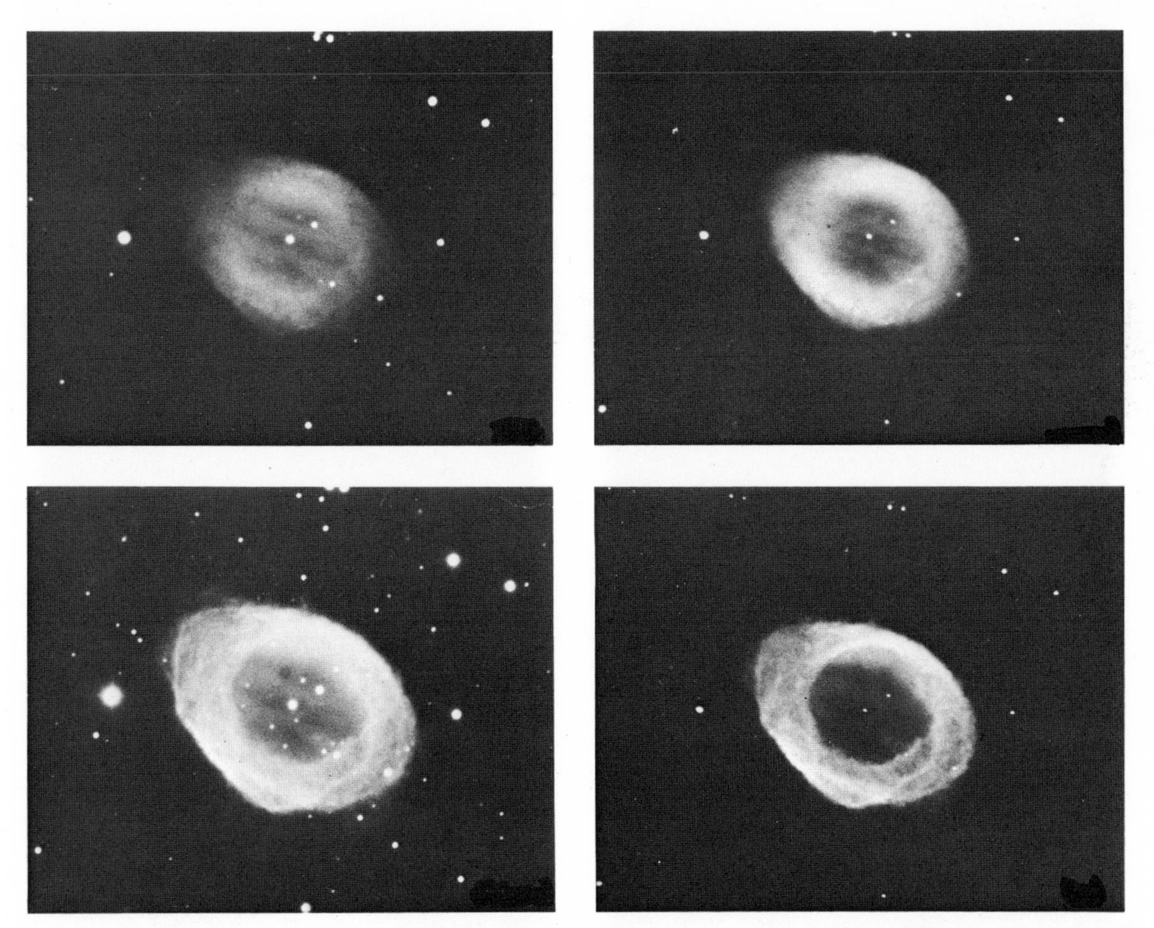

M57, the Ring Nebula in Lyra, shows striking differences in size and structure when photographed in blue (upper left), green (upper right), yellow (lower left), and red (lower right) light. The stars, with the exception of the one at the center of the ring, are all field objects.
—MOUNT WILSON AND PALOMAR OBSERVATORIES

M1, the Crab Nebula, photographed in red light.
—Mount Wilson and Palomar Observatories

lines due to hydrogen, neutral and ionized helium, and ionized nitrogen, oxygen, and sulfur, which have their origin mainly in the filaments.

As mentioned earlier, the Crab Nebula is the remains of the supernova of 1054. In the far distant future perhaps, it may assume the ringlike appearance of many planetaries. Spectroscopic observations show that the gases are expanding at a velocity of about 700 miles a second. This rate, considered along with a slight increase in the nebula's apparent size over the last twenty years, enables us to put its distance at 3,500 to 4,000 light-years. Chinese records state that it remained visible for two years and could be seen in daylight for twenty-three days, so at its maximum it was at least as bright as Venus. Its luminosity must therefore have reached the fantastic intensity of about 350 million suns.

The Veil Nebula Photographs of parts of the Milky Way taken with rapid-lens, wide-angle sky cameras reveal numerous curved filaments of veil-like nebulosity arranged in the form of rings. Among the most striking is the Veil Nebula, or Network Nebula, in Cygnus. Part of it, NGC 6960, involves the fourth-magnitude star 52 Cygni and was described by Herschel as "a milky Ray." He also detected two other parts (NGC 6992–6995), but there was nothing to suggest that they were associated with NGC 6960. Small-scale photographs taken in the red light of hydrogen show that these filaments, along with others, form part of a remarkable loop- or ring-shaped structure nearly 3 degrees in diameter. Nor is this all. Proper-motion estimates based on photographs taken many years apart make it clear that the loop is expanding, and spectroscopic observations put the rate of expansion of the part nearest us at roughly 60 miles a second. The available data give the loop a distance of 2,000 light-years and a diameter of about 120 light-years.

The Veil Nebula is an enormous hollow bubble of hydrogen gas. It can emit no light of its own and is probably subject to the ionizing radiation of a hot central star. The field in which the nebula lies is extremely rich in faint stars, but despite careful searches no suitable candidate has been found. The general opinion is that the nebula is the remnant of a supernova outburst, and this is supported by the fact that the area is also a source of radio waves. On this assumption, together with another which supposes a steady decrease in the rate of expansion, the explosion took place between 10,000 and 20,000 years ago.

CASSIOPEIA A Loop nebulae and parts of loop nebulae have also been found in other parts of the Milky Way. One in Cassiopeia coincides with a particularly strong source of radio emission known as "Cassiopeia A" but is about 10 degrees away from the site of the supernova of 1572. It consists of a group of small fragments of bright nebulosity whose radial velocities, of the order of 2,500 miles a second, indicate that they form part of a rapidly expanding residual supernova shell. Their distance from the earth is thought to be of the order of 11,000 light-years.

Another loop nebula, known as S147 and

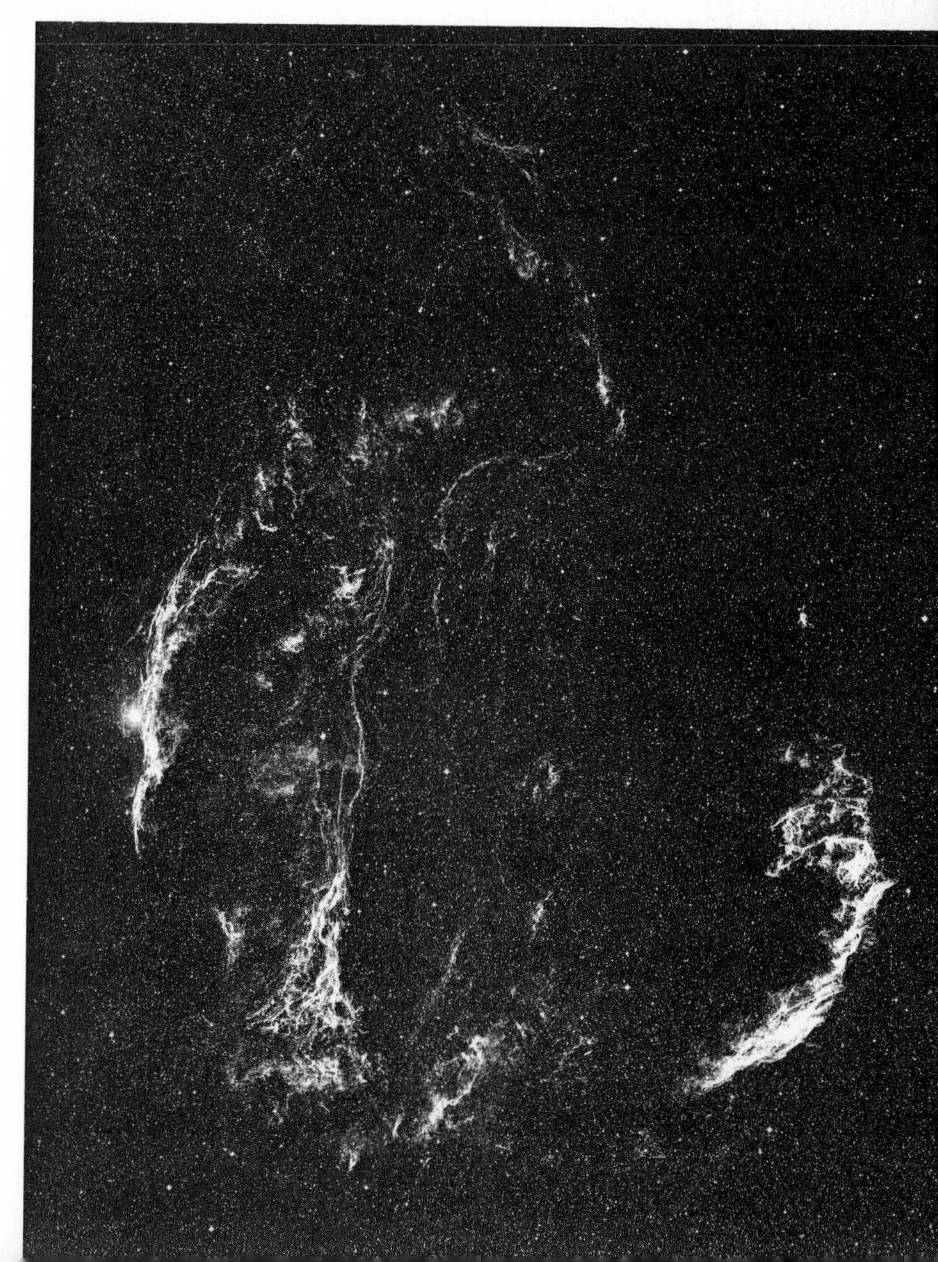

The Veil Nebula in Cygnus. This small-scale photograph in red light shows the whole area of nebulosity. The star 52 Cygni and NGC 6960 are on the extreme left.
—MOUNT WILSON AND PALOMAR OBSERVATORIES

located in Auriga, was discovered in 1952 by G. A. Shajn and Miss V. F. Hase of the Crimean Astrophysical Observatory. They described it as "a network of exceedingly narrow filaments forming many intersecting arcs" and gave it dimensions of roughly 3 by 2 degrees. It has a highly intricate threadlike structure, is a source of radio waves, and is believed to be between 2,500 and 3,500 light-years distant. It lies near Nova Aurigae 1891, but the two objects are considered to be unrelated.

NATURE OF LOOP NEBULAE The rapidly expanding gases and associated shock wave of a stellar explosion do not necessarily have unhindered motion through space. Their speed can be effectively reduced by the interstellar medium, which consists of an extremely tenuous gas and its condensations, together with rarefied clouds of dust. This material is concentrated in and near the general plane of the galaxy, and it is precisely in this region that loop nebulae are found. The Veil Nebula in Cygnus, for instance, lies in the Milky Way and has a velocity of expansion well below that of a supernova outburst in its early stages. So if it is the result of a supernova explosion its motion has presumably been slowed down by the resisting medium. It has also been suggested that collisions between parts of the shell and dust clouds might account for the veil-like structure of nebulous filaments. Also, that these collisions are responsible for part if not all of the filament radiation in both optical and radio wavelengths. But although a great deal of theoretical work has been done on the subject, the factors involved continue to remain both numerous and perplexing.

"Holes in the Heavens"

In many parts of the Milky Way intervening dust clouds effectively hide all the stars that happen to lie behind them. Their existence was first noticed by William Herschel when he came across an apparent "hole in the heavens" slightly north of the star 19 Scorpii. Nearby was the globular cluster M80, whose brightness made the dark patch seem even darker. It was as if M80 had drawn off all the stars from that place, thereby leaving a vacancy. John Herschel came across nearly fifty similar "holes" or "dark vacancies" in the Southern Hemisphere. One of them, near the Southern Cross, covered about 40 square degrees of the Milky Way. Known as the Coalsack, it is readily visible to the unaided eye and was doubtless known to navigators in the sixteenth century. Early observers described it variously as a hole, an opening in the heavens, an inky black cloud, or a vacancy in the Milky Way.

IDENTIFICATION OF DARK NEBULAE Most astronomers of the last century thought that the Coalsack and other dark areas arose from intervals of starless space—from a real paucity of stars compared with the richness of neighboring regions. Their true character was recognized about 1903 by Max Wolf of Heidelberg. Using a 16-inch short-focus sky camera he obtained a large number of photographs of nebulae in the Milky Way, among them one of NGC 7000, or the North America Nebula in Cygnus. The nebula was surrounded by a dark "halo" or "trench" so distinct and empty of faint stars as to force the conclusion that it was a dark cloud.

Wolf's surmise was confirmed by E. E. Barnard, another pioneer in modern astronomical photography. Barnard's photographic survey of the Milky Way, begun at the Lick Observatory in 1889, showed with insistent clarity that the dark patches were dark nebulae, or clouds of obscuring material. The survey continued through to 1927, when the publication of Barnard's magnificent *Atlas of Selected Regions of the Milky Way* provided astronomers with the first permanent record of the number and distribution of dark nebulae.

GIANT DARK NEBULAE Some dark nebulae are much larger than the Coalsack. In Ophiuchus, for example, a dark nebula covers an enormous area of about one thousand square degrees, but the large number of foreground stars dotted over it renders most of it relatively inconspicuous. This giant, together with similar objects in nearby Scorpius, is apparently a member of a complex and comparatively near group of dark nebulae. Some of its extensions, particularly those in the region of Rho and Theta

The dark nebula Barnard 72 in Ophiuchus
—MOUNT WILSON OBSERVATORY

Ophiuchi, look like long dark lanes smudged out among the stars. Another large and complex group is found in the regions of Perseus, Taurus, and Orion, while a third, in Cygnus and Serpens, forms a dark rift which divides the Milky Way into two branches.

The appearance of a dark nebula is obviously dictated by its size, distance, and opacity.. The blackest are those which, being comparatively near and opaque, blot out the light of all the stars beyond. Those at greater distances have foreground stars projected on them, while the extremely remote ones are so covered with stars as to be relatively inconspicuous. The traceable forms of dark nebulae reveal next to nothing about their actual sizes and powers of absorption.

In many cases partial "breaks" in the dark clouds enable some light from the regions beyond to get through. The Coalsack, for instance, is not completely dark but contains many faint stars and several flecks of nebulosity indicative of bright gaseous regions beyond. Stars seen through fairly thin or rarefied dust clouds appear dimmed and redder than they should be. The effect is similar to that produced by the earth's atmosphere when the sun is about to set. Sunlight has then to pass through a great thickness of air molecules and dust particles. In the process it is selectively scattered: the short-wave components are scattered to produce the blue-sky effect but the longer waves pass through relatively undisturbed. As a result the sun looks like a red M-type star, although its spectrum shows that it is really a G2-type star. In a similar way many B-type stars, like Zeta Persei and 55 Cygni, have color indices which correspond to G-type stars. They can thus be used to obtain a rough estimate of the amount of absorption, and indirectly, some idea of the nature of the obscuring medium.

Reflection Nebulae

When dark clouds are seen by the light they scatter they are referred to as reflection nebulae. They can also reflect the light of nearby stars, but the term "reflection" as generally used implies both reflection and scattering. The incident starlight is spread over an immense area, so while reflection nebulae are faint objects their material must be an excellent light reflector. All the available evidence indicates that they are clouds of icelike grains rather than clouds of meteoritic dust. The grains probably consist of frozen compounds of hydrogen, oxygen, and carbon and are generally much larger than atoms and molecules. They are certainly not dust particles, although polarization studies suggest that they contain small percentages of iron and other metals. Grains that happened to come within the effective range of a hot B-type star would be evaporated or swept away by radiation pressure, but those at greater distances would merely reflect and scatter the star's light.

Prominent Associations of Bright and Dark Nebulae

Dark nebulae are mainly responsible for the highly intricate and irregular shapes of most bright nebulae. Where their parts overlay or encroach on bright nebulosity they stand out as dark lanes, filaments, and wisps. Conspicuous examples are the nebulae around Eta Carinae and Gamma Cygni, the Great Nebula in Orion, the Lagoon Nebula (M8 in Sagittarius), the Trifid Nebula (M20 in Sagittarius), the Rosette Nebula in Monoceros, and M17 in Scutum. Of these the Eta Carinae nebula, sometimes called the Keyhole Nebula, is one of the largest and most beautiful. On photographs it can be traced over an area at least 25 times the apparent area of the full moon. Eta Carinae itself is a nova-type variable surrounded by an extensive gaseous halo whose spectrum becomes visible only when the star is at its faintest. Since 1865 the star has been fairly constant near the eighth magnitude, but between 1835 and 1843 it rose from the fourth magnitude, to rival Canopus, the second brightest star in the night sky. Meanwhile the nebula showed no perceptible change in brightness; it owes its luminescence not to Eta Carinae but to numerous hot O- and B-type stars immersed in it.

M20, the Trifid Nebula in Sagittarius. Photographed in red light.
—MOUNT WILSON AND PALOMAR OBSERVATORIES

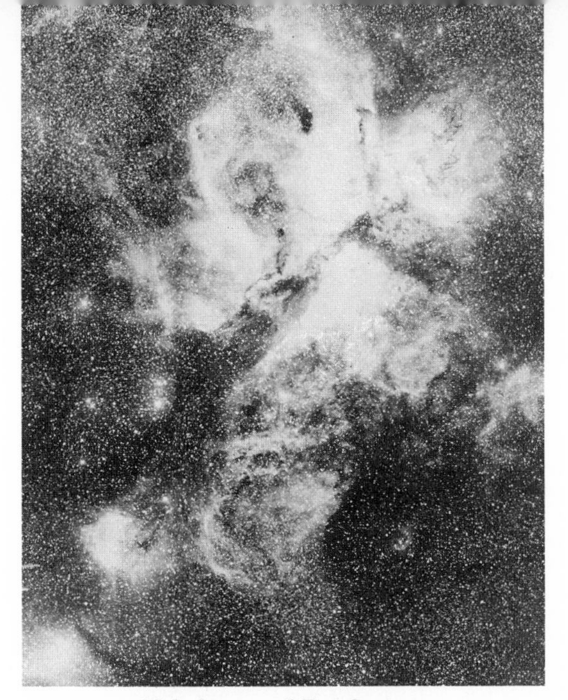

Nebula around Eta Carinae.
—Mount Stromlo Observatory

Globules In the case of M8, M16, the Eta Carinae nebula, the Rosette Nebula, and several others, high-resolution photographs taken with large telescopes and red-sensitive plates reveal numerous small, roundish, dark nebulae projected on a bright background. B. J. Bok has referred to them as "globules," for they are presumably ball-shaped clouds of dust with masses of the same order as those of stars, and diameters that range from about 0.1 light-year to 20 to 25 light-years. He has suggested that they may be "protostars," or objects in the process of contracting into stars. The Coalsack, with a mass between 25 and 50 times that of the sun, is probably a giant globule. Its distance of approximately 500 light-years makes it one of the nearer dark nebulae and puts its diameter at approximately 50 light-years.

The Great Nebula in Orion The Eta Carinae nebula has a declination of about −59 degrees and is therefore too far south to be visible from mid-northern latitudes. In recompense, northern observers have the Great Nebula in Orion, a most conspicuous and majestic object situated in a particularly interesting part of the sky. On a clear dark night it is readily visible to the unaided eye, but there is no record that it was ever noticed before the introduction of the telescope. It was seen in 1618 by J. B. Cysat of Ingolstadt and first described in 1659 by the Dutch astronomer C. Huygens:

An enlarged section of NGC 2237, the Rosette Nebula in Monoceros, photographed in red light and showing numerous globules projected against the luminous background.
—Mount Wilson and Palomar Observatories

In the Sword of Orion, are three stars quite close together. In 1656, as I chanced to be viewing the middle one of these with the telescope, instead of a single star, twelve showed themselves [a not uncommon circumstance]. Three of these almost touched each other, and with four others, shone through a nebula so that the space around them seemed brighter than the rest of the heavens which was entirely clear and appeared quite black, the effect being that of an opening in the sky through which a brighter region was visible.

The Trapezium Messier saw the nebula as two adjacent masses and listed them as items 42 and 43 in his catalog, but in reality one mass merges into the other to form a single complex.

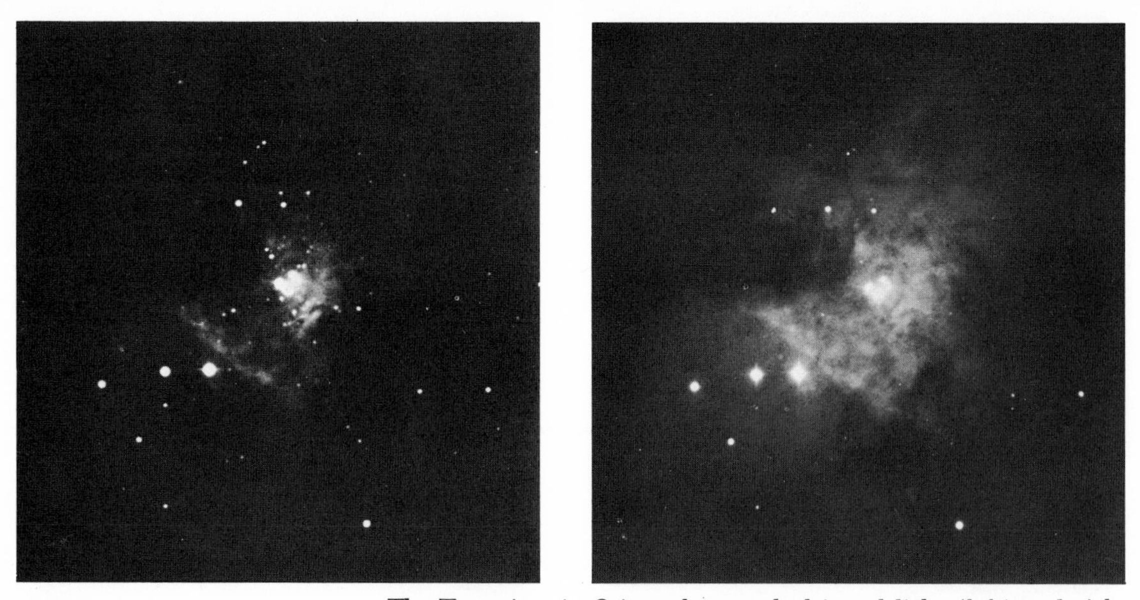

The Trapezium in Orion, photographed in red light (left) and violet light (right).

—LICK OBSERVATORY

The brightest parts are concentrated on a group of stars known as Theta Orionis. Four of these form the subgroup Theta-One Orionis, commonly called the Trapezium. These stars, all spectroscopic binaries with very hot components, are the main exciting agents. Together they spread their ionizing effect over a nebulous mass at least 1 degree across, or twice the apparent diameter of the full moon. Since the nebula has a distance of about 1,500 light-years this corresponds to a linear size of not less than 20 light-years.

The Trapezium is best seen visually in the telescope or on short-exposure photographs. With increasing exposure times the stars become lost in the brightness of their surroundings and eventually the whole central region looks no more than a bright amorphous mass. Recent studies by K. Strand and others show that the stars of the Trapezium are not isolated objects but belong to a compact cluster of at least three hundred stars. Part of the cluster, along with some of the surrounding nebulosity, is hidden behind clouds of interstellar dust, but if these were not present more than four hundred stars might become visible.

Intervening dust clouds undoubtedly account for the numerous dark rifts and blotches that give the nebula such a chaotic appearance, and also for the Fish's-Mouth, a dark indentation in the northeastern boundary. Fortunately, the dust is transparent to radio waves, which are also emitted by the nebula itself, so it has been possible to trace the nebula's actual outline. This, as expected, is roughly circular, and its central region of highest intensity coincides with the Trapezium cluster. The nebula is a vast spherical mass of luminescent gas of extremely low average density.

Proper-motion studies by P. P. Parenago and Strand show that the Trapezium cluster is expanding, and at such a rate as to indicate a common origin for its stars sometime between 10,000 and 300,000 years ago. The nebula too has an overall expansion, despite various turbulent motions within its mass. Different regions expand with different velocities, but the general rate, some 5 or 6 miles a second, is comparatively small. It has been suggested that the cluster stars have just emerged from the cloud of interstellar gas in which they had their birth. If this is so the nebula represents the residual gas of recent star formation and shares a common ancestry with the cluster. All the available evidence subscribes to this view and to an estimated age of less than 50,000 years.

THE EXTENT OF THE GREAT NEBULA IN ORION The spectrum of the Great Nebula in Orion consists of a number of bright emission lines superimposed on a faint continuous background. The bright lines, as Bowen discovered, are due to several elements in various stages of ionization. The process of light emission is similar to that operating in planetary nebulae. Once again hydrogen is by far the most abundant element. The faint continuous background arises from reflection nebulosity. Radiation from the exciting stars and the nebula itself is reflected by dust clouds associated with the gaseous masses. The great problem here, as with other complex irregular nebulae, is to give their two-dimensional photographs a three-dimensional interpretation. In this case, intervening dust clouds obscure much of the nebula, but faintly reflecting material involved in the nebulosity reinforces its light. Huygens' description of the nebula as "an opening in the sky through which a brighter region was visible" was reasonably apt.

The Great Nebula is the brightest part of a vast complex of gas and dust that involves the entire constellation of Orion. Immediately to its north lies the nebulous patch NGC 1977, partly hidden by dark material. Much farther north lies the visual binary Alnitak, or Zeta Orionis, the most southerly of the three bright stars in Orion's belt and the source of excitation of another bright nebulous mass. Here, IC 434, a dark nebula with a luminous fringe, forms a remarkable outline shaped like a horse's head. Bright fringes of this type, looking like the "silver linings" of terrestrial storm clouds, frequently border dark nebulae. They are thought to be the results of impacts between expanding masses of hydrogen gas and the borders of dark nebulae.

Stellar Associations

Many of the stars in the region of Orion's Belt and Sword form rich but loose groupings. The stars involved are hot, highly luminous O- and B-type objects, widely spaced but, it is thought, physically associated. Groupings of this kind are found in Perseus, Lacerta, Scorpius, Centaurus, and other regions rich in nebular ma-

The Great Nebula in Orion, M42 and M43.
—Lick Observatory

IC 434, the Horsehead Nebula, a cloud of obscuring dust south of Zeta Orionis. Photographed in red light.
—Mount Wilson and Palomar Observatories

terial. The groupings, called stellar associations have been extensively studied by V. A. Ambarzumian and his colleagues in the Soviet Union and by W. W. Morgan, G. Munch, and others at the Yerkes Observatory. The stars of any one association are thought to have had a common origin and a similar past history. Their distribution in space and their involvement with clouds of hydrogen suggest that they are no more than a few million years old and therefore mere infants in the stellar community. One particular association, Perseus II, is expanding, and this may be true of the association clustered around the center of Orion. At least three stars, AE Aurigae, Mu Columbae, and 53 Arietis are moving away from this region at speeds which suggest that they had their origin there some 3 million years ago.

T TAURI STARS Orion also contains several examples of what Ambarzumian has called "T associations." Their member stars, all faint and irregularly variable in their light, have spectral peculiarities similar to those of T Tauri, a dwarf G5 star whose spectrum is crossed by strong emission lines of hydrogen and a few ionized metals. They are invariably associated with dark nebulosity, as for example in the regions of Auriga, Taurus, Orion, Ophiuchus, Monoceros, and Corona Australis. Most of them have underlying spectra between types A and M, and all are dwarfs. Their brightness range is generally of the order of one or two magnitudes, but among individuals the light behavior is extremely diverse. Some undergo sudden flarelike surges, others have small but rapid fluctuations, but the changes are always erratic and therefore quite unpredictable.

Ambarzumian and others have suggested that T Tauri stars are very young objects recently formed from clouds of gas and dust. Their tendency to fall above the main sequence and to be in association with interstellar clouds and also with hot, young stars (as in the region of the Great Nebula in Orion) favor the view that they are contracting gas spheres which have not yet reached the main sequence. Alternatively, they may be normal stars that have wandered into regions rich in nebular material. The infall of this material might then trigger off flares or build up bright and extensive "chromospheres." But whatever the explanation it is significant that moderately cool dwarfs of spectral types F, G, and M seem to be much more affected by surrounding nebulous material than are hot dwarfs of spectral types O, B, and A.

Variable Nebulae

Most of our present knowledge of T Tauri stars is due to the work of three astronomers, Joy at the Mount Wilson and Palomar Observatories, G. H. Herbig at Lick Observatory, and G. Haro at Tonantzintla Observatory, Mexico. In particular, Herbig and Haro have made a close study of T Tauri stars associated with variable nebulae.

One of these is T Tauri itself, for as J. R. Hind discovered in 1852, it lies quite close to the small nebula NGC 1555. Hind saw the nebula in a telescope of modest size, but between 1864 and 1890 it disappeared completely even in large instruments. It then reappeared and has since gradually increased in brightness. The star is also associated with a small condensed variable nebula which immediately surrounds it, and with a third variable nebula, NGC 1554, some 4 minutes of arc distant.

Two other interesting examples are NGC 2261 in Monoceros and NGC 6729 in Corona Australis. Both have a cometary or fanlike appearance, with the T Tauri-type stars R Monocerotis and R Coronae Australis respectively at their tips, or "heads." William Herschel described NGC 2261 as "a star with nebulosity attached," but its variability was discovered much later by E. Hubble of the Mount Wilson Observatory. Hubble found that bright knots and wisps in the fan disappeared only to reappear many years later. On one occasion they faded and brightened in only a few days as though a dark shadow were sweeping across the fan. The fan would therefore appear to be a fairly static reflection nebula whose variability arises from dark material passing between it and the illuminating star.

A fourth example is IC 405, a variable nebula associated with AE Aurigae. The latter, as already mentioned is apparently a member of an Orion association of hot O- and B-type stars. Herbig's observations indicate that the star has encountered two separate nebulae. One, a luminescent gas cloud, is being ionized by AE Aurigae, while the other, composed of interstellar dust, shines by reflected and/or scattered light.

Herbig-Haro Objects

Herbig and Haro have also given close attention to a number of extremely faint objects which are a kind of cross between "cloudy stars" and "stars with nebulosity attached." Each consists of a small emission nebula with a starlike nucleus. In this respect they resemble T Tauri and the small condensed emission nebula surrounding it. Their most remarkable feature is revealed by photographs taken at intervals of several years. These show that some objects have developed further "nuclei," but whether the latter are embryo stars formed by fission or condensation, or whether they are even stars at all, has still to be determined. If the "nuclei" are newborn stars the Harbig-Haro objects may be a direct link between dark globules and T Tauri stars.

Radio Emissions from Nebulae

Many gaseous nebulae are sources of radio energy, the character of which points to two different types of origin. That emitted by the Crab Nebula and other supernova remnants is said to be nonthermal since it is attributable to the synchrotron mechanism. That emitted by the Great Nebula in Orion and other diffuse gaseous nebulae is thermal. It gives rise to a continuous radio spectrum covering a wide range of frequency and comes from hot, ionized hydrogen (HII) near the exciting stars. Also classed as thermal is the radiation from cool neutral atomic hydrogen (HI) outside the range of ionization. This is characterized by its specific frequency of about 1,420 megacycles a second, equivalent to a wavelength of approximately 21 centimeters.

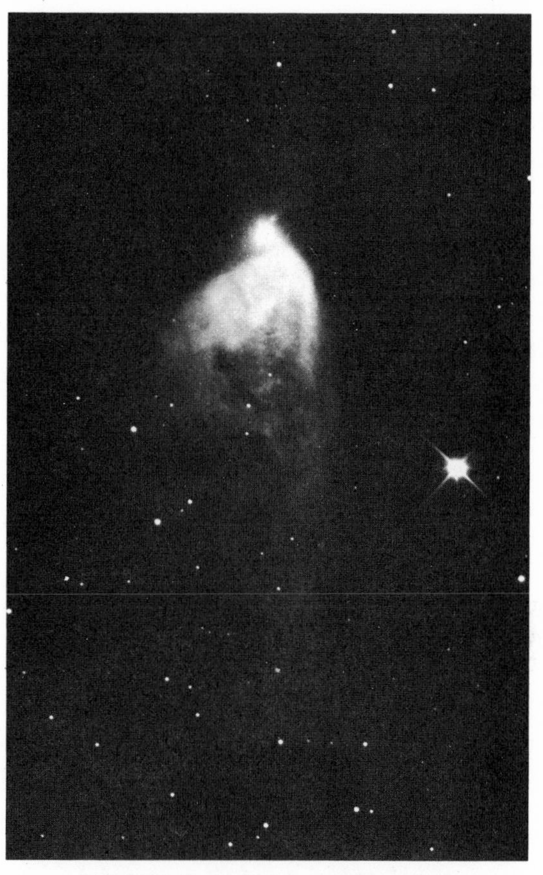

NGC 2261, Hubble's variable nebula.
—LICK OBSERVATORY

One of the Herbig-Haro objects.
—LICK OBSERVATORY

The Interstellar Medium

The distribution of HII and HI can be determined by using radio telescopes of sufficient resolving power "tuned" to receive particular frequencies. The work done so far amply confirms an earlier surmise that bright diffuse nebulae are the visible parts of an extensive and highly complex interstellar medium. The medium consists largely of hydrogen and has its greatest concentration in the plane of the galaxy. In particular, measurements of displacements of the 21-centimeter line due to the Doppler effect provide valuable information about the motions of HI clouds in the line of sight, while measures of its profile give an insight into their densities and temperatures. As a result it has been possible to probe regions otherwise hidden by obscuring dust clouds and to discover that the hydrogen clouds (and hence diffuse gaseous nebulae) are concentrated in the spiral arms of the galaxy.

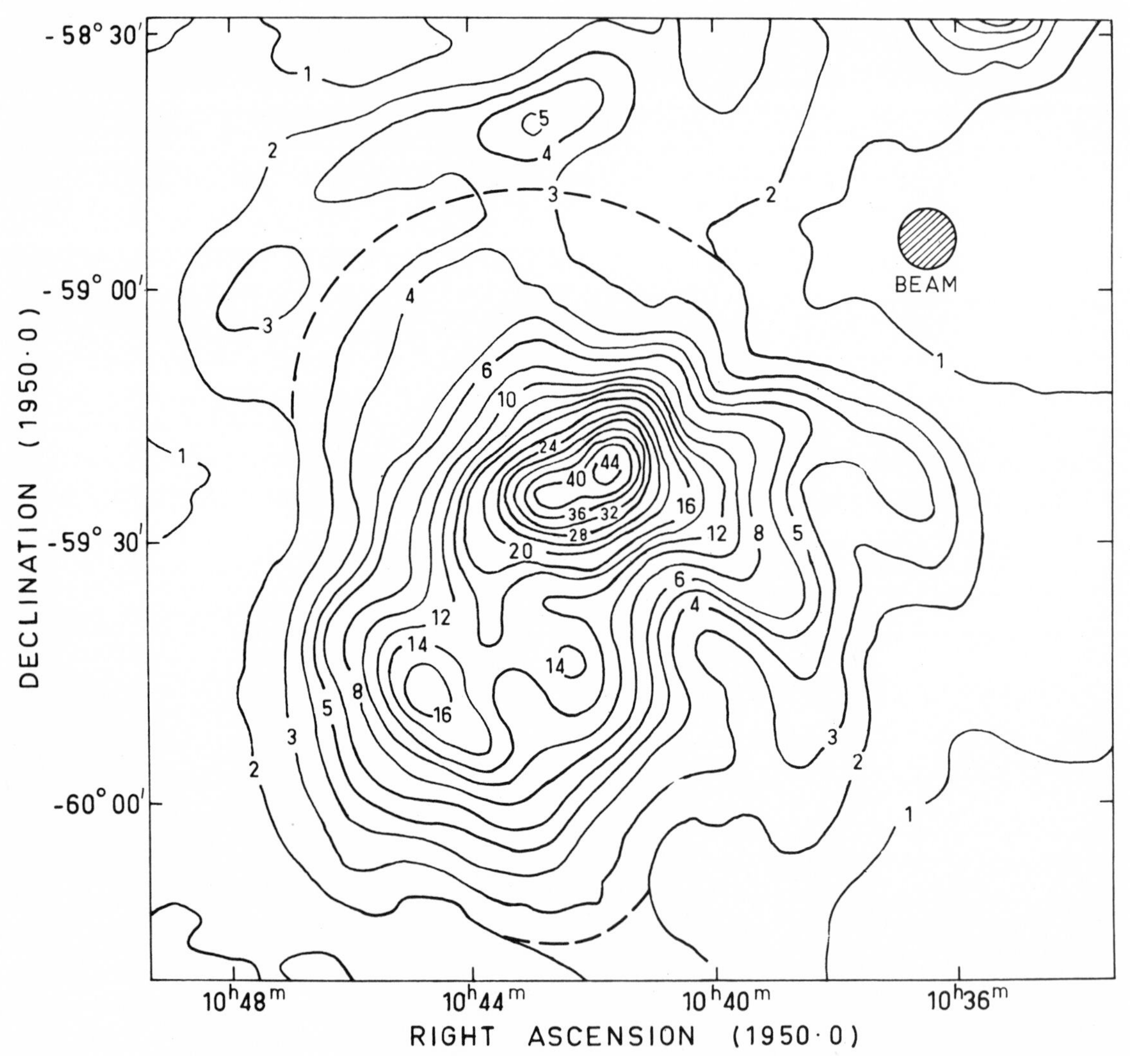

Contours of relative intensity at a wavelength of 11.3 centimeters in the region of the nebula around Eta Carinae, as recorded by the 210-foot radio telescope at Parkes, New South Wales, Australia.
—DIVISION OF RADIOPHYSICS, CSIRO, SYDNEY, AUSTRALIA

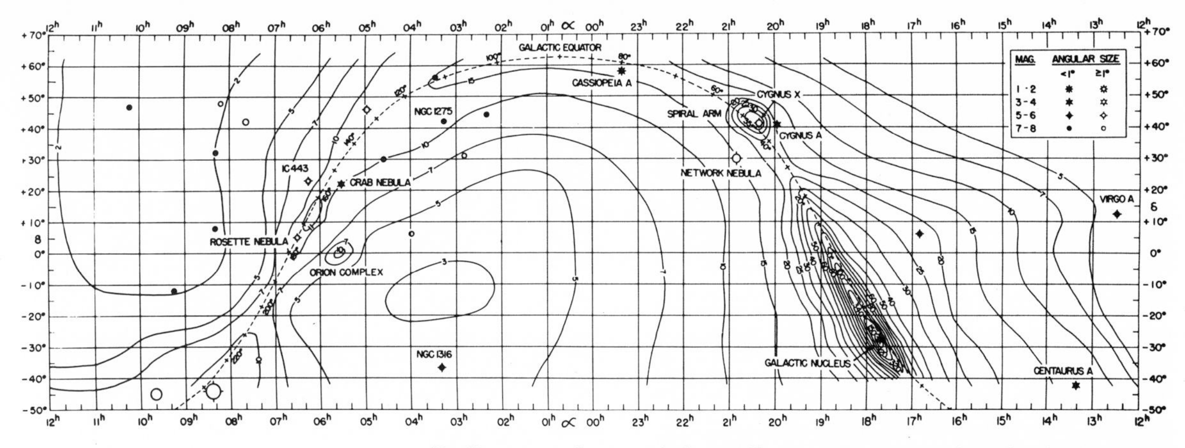

Radio map of part of the Milky Way at a wavelength of 12 meters.
—Ohio State–Ohio Wesleyan Radio Observatory

CHAPTER X

STAR CLUSTERS

Since the sun, and every other star, is moving through space, familiar star groups and patterns in our skies are slowly changing in appearance. For most stars the shift in relative position is extremely small, even in a century, but the effect is cumulative and can lead to noticeable results in the course of many thousand years. The changes are not entirely random. The sun's motion alone insures that other stars have a common apparent motion towards the region of Canis Major and away from Lyra. Then there are the common motions characteristic of stellar associations. In Perseus, for example, a number of hot O- and B-type stars form an expanding group or association, while many bright stars in Scorpius, Lupus, and Centaurus travel along together in slightly divergent directions. In addition there are several moving "open" or "loose" clusters, like the Hyades and Pleiades, whose stars belong together and share a common space motion.

Moving Systems of Stars

The suspicion that the stars of certain groups might have the same proper motion was entertained by Bessel, Schönfeld, R. Proctor, and several others well before the end of the last century. Proctor even went to the trouble of comparing the positions of stars in various parts of the sky with those given in early catalogs, and in 1869, he obtained evidence of what he called "stardrift." Five of the seven stars of the Big Dipper, for example, were found to be drifting "nearly in the same direction and at nearly the same rate." Later studies, based on radial velocity estimates and more accurate proper motion determinations, not only confirmed this but showed the communal drifting extended to include Sirius and at least forty other stars. All these stars are members of a moving system known as the Ursa Major Cluster. They are widely scattered over the sky because the cluster is both open and very near to us. For the same reason it includes a large number of non-cluster or field stars, among them the two endmost stars of the Big Dipper (Alpha and Eta Ursae Majoris) and even the sun.

The Hyades

Proctor was also able to obtain evidence of a general drifting of several stars in the region of the Hyades; but it was left to the American astronomer Lewis Boss, some forty years later, to show that some forty stars of the Hyades were physically associated. The Hyades is a conspicuous open cluster in Taurus. To the unaided eye the brighter components form a V-shaped pattern (the face of the Bull), which includes orange-red Aldebaran. This group, however, is only the central part of a cluster of at least 150 stars (and probably as many as 350) that covers a roughly circular area some 25 degrees across. Aldebaran is a field star, being about 65 light-years away, whereas the cluster itself lies at twice this distance. Its space motion, moreover, differs from that of the other Hyades, and on this account alone it cannot be a member of the cluster.

The radial velocities and proper motions of the stars of the Hyades have received a great deal of attention because they provide reliable estimates of distances. They indicate that the center of the cluster is about 120 light-years away, that its overall diameter is about 60 light-years, and that there is no appreciable rotation or expansion of the cluster as a whole. Most important of all, they reveal that the cluster stars are moving away from us with equal velocities in parallel paths.

The average radial velocity is about +24 miles a second (the plus sign indicating velocity of recession) and the proper motions all converge to a point located a few degrees to the east of Betelgeuse. The cluster is analogous to a flock of birds flying in fairly close formation. As the flock recedes it appears to slow down and get smaller until it eventually becomes a speck on the skyline. The Hyades have a space velocity of about 27 miles a second, but the distance they have to travel is so great that their formation will not begin to look appreciably smaller until many thousand years have passed. Calculations by Boss some years ago showed that the brighter Hyades would be ninth-magnitude stars some 65 million years hence. The cluster will then look like many other present-day telescopic clusters—small, faint, fairly well-condensed, and relatively insignificant.

Most of the stars of the Hyades lie on the main sequence between spectral types K5 and A5. They form the backbone of the cluster and include stars almost identical to the sun. Some of the brighter stars lie above the main sequence. They include Theta-2 Tauri, a white giant of spectral type A7 and the brightest member, also a few giants of types G and K. Now since these stars belong together, it is reasonable to suppose that they emerged from one and the same gas-and-dust cloud in a comparatively short period of time. Just as originally they had roughly the same chemical composition, so now they have approximately the same age. Their present differences could be due largely to differences in their initial masses. The giants, no older than the dwarfs, have eaten deeper into their nuclear resources in order to

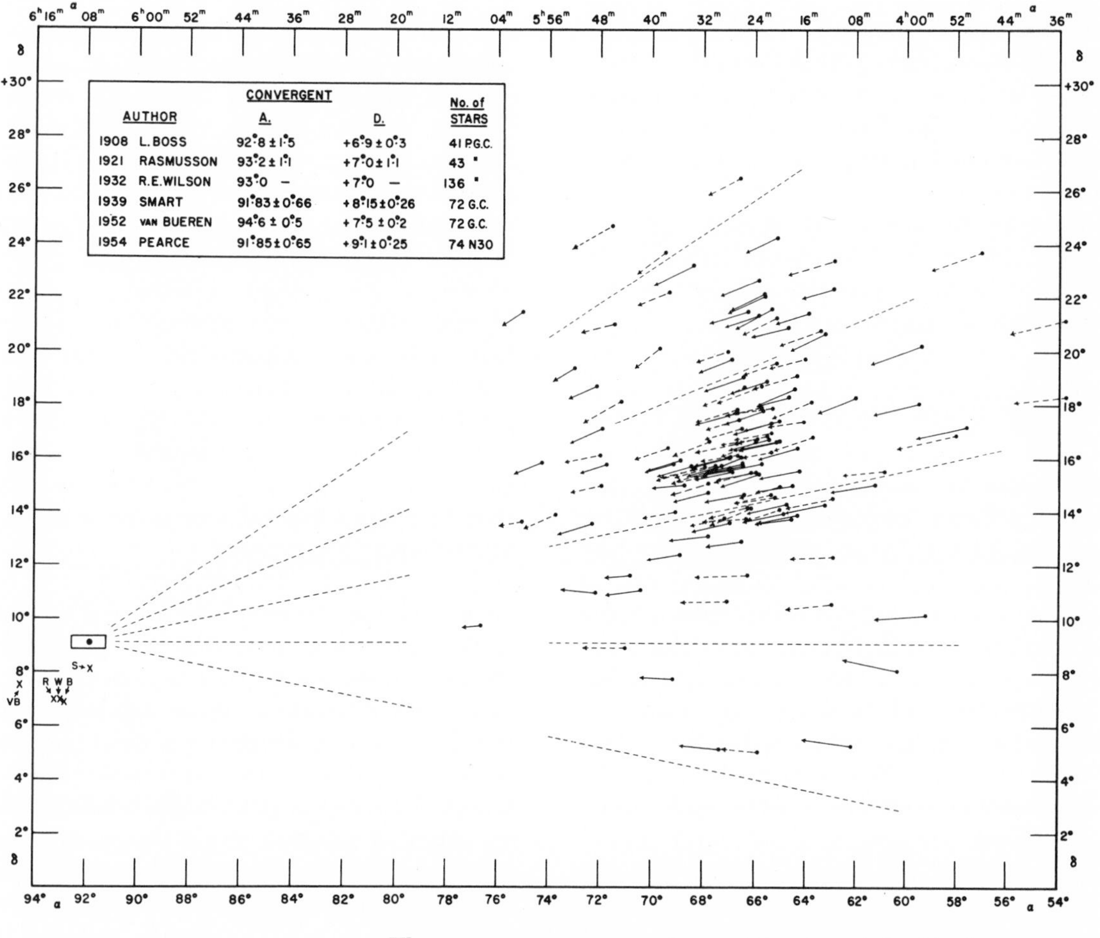

The convergent proper motions of the stars of the Hyades.
—DOMINION ASTROPHYSICAL OBSERVATORY, VICTORIA, B.C., CANADA

maintain a high level of energy output. They have, so to speak, "lived it up," but at the expense of aging more quickly. In the far distant future, of course, the cluster will contain many more giants, but the increase in its overall luminosity will probably be more than offset by its increased distance from us.

The Pleiades

A group that is even more conspicuous than the Hyades is that known as the Pleiades, also in Taurus. Its central part covers an area about equal to that of the full moon and its brightest stars lie roughly between the third and sixth magnitudes. Alcyone, the brightest star, is of the third magnitude. Next in brightness comes Electra and Atlas (magnitude 3.6), Maia (magnitude 3.9), Merope and Taygeta (between the fourth and fifth magnitudes) and Pleione (magnitude 5.1). The first six of these can be seen with the unaided eye without much difficulty, but with good vision and a clear, dark night, the number can be increased to as many as fourteen.

In the Hyades the most massive stars tend to be concentrated towards the center of the cluster. A similar tendency is found in the Pleiades. The central part is about 400 light-years away, so any main-sequence stars in the cluster will be more than nine times fainter than corresponding main-sequence stars in the Hyades. At this distance the sun would appear no brighter than a star of magnitude 10.5. About 250 stars have been recognized as cluster members, but there are probably many more. Most of them fall near the main sequence, but compared with the Hyades, they have a much greater range in brightness and mass. The brightest are not yellow or orange giants but very hot and rapidly evolving B-type stars well over a thousand times more luminous than the sun. The faintest are K-type dwarfs, but there are doubtless many more faint dwarfs as yet undetected.

An H-R diagram for the Pleiades shows two highly significant features. Many of the stars, especially those of spectral types G, K, and M, lie slightly above the main sequence. This is readily explained if we regard them as young, presequence objects in the final stages of contraction. The hot B-type members also lie above the main sequence, where their excessive luminosities proclaim them to be stars that are moving from dwarf to giant status. If these interpretations are correct the cluster is much younger than the Hyades. Theoretical calculations indicate that it has taken about 60 million years to reach its present state, whereas the Hyades has taken between 1 and 2 billion years.

The Pleiades declares its comparative youth in another way. The central knot of bright massive stars is embedded in nebulosity, first detected in 1859 by W. Tempel, who noticed that the star Merope had a misty appendage shaped like a comet's tail. Later observers found misty shreds and patches near Maia and other members of the group, but only dry-plate photography could reveal the full extent and extreme complexity of the involvement. The nebulosity is a first-class example of a reflection nebula, the first of its kind to be discovered. In 1912 V. M. Slipher at Lowell Observatory found that its spectrum was similar to that of the involved stars, so it was obviously made visible by reflected and scattered light.

Radio observations indicate that the Pleiades also contains interstellar gas in the form of neutral hydrogen, but its full extent and the degree of concentration are not yet known. The radiation may be the result of interaction between the cluster and the interstellar gas through which it is moving. On the other hand the gas may be moving with the cluster and represent part of the parent gas cloud in which the Pleiades was born.

Be-type stars are remarkable for their high speeds of rotation. At least six of the brighter Pleiades come in this category. Their spectra contain very broad and diffuse absorption lines coupled with emission features which indicate speeds of rotation of the order of 200 miles a second. Maia is probably a seventh example; its lines are fairly narrow but it could still be a rapidly rotating star whose axis happens to be inclined at a small angle to the line of sight.

Pleione is the prototype of all so-called shell stars. In its spectrum each broadened hydrogen-

Reflection nebulosity in the Pleiades.
—LICK OBSERVATORY

absorption line has a narrow dark line, or "core," at its center. The core is flanked on both sides by narrow emission lines. These peculiarities suggest that the star is surrounded by a broad but extremely diffuse gaseous ring or shell. The narrow cores, along with narrow dark lines elsewhere in the spectrum, are due to absorption by that part of the ring or shell which intercepts the light of the star. The emission lines arise from the bright "wings" of the ring, that is, from those parts which do not intercept the light of the star.

It is tempting to imagine that the ring was shed or is being shed as a direct result of the star's high speed of rotation. Yet, strange to tell, the emission lines, and presumably the ring, disappeared altogether between 1906 and 1938. Since 1938 the spectrum has shown its earlier emission features, but at first the absorption lines slowly increased in intensity and then slowly decreased. These and other changes in spectrum indicate that Pleione has shed another ring. The loss in mass, however, and the consequent enrichment of the interstellar medium, must be very small—smaller even than that involved in a nova outburst.

Greenstein and Eggen of the Mount Wilson and Palomar Observatories have recently drawn attention to no less than fourteen white dwarfs in the Hyades and at least one in the Pleiades. If modern evolutionary theory is correct, the presence of these objects in fairly young clusters means that some very massive stars have evolved with great rapidity. The one in the Pleiades must have raced through the greater part of its life at a phenomenal speed. Of course, the assumption that all the stars of the Pleiades are of the same age may not be valid. The process of star formation could be spread over a million years or more. If this were so its effect would be negligible in an old cluster but quite noticeable in a young one. Either way, we still have to face the problem of how a massive star can lose enough of its mass to become a white dwarf.

Other Naked-eye Star Clusters

We have dealt with the Hyades and Pleiades at some length since they are familiar objects and also because they have been extensively measured and studied. Other star clusters, apart from Coma Berenices and NGC 2516, are more distant and therefore fainter, and this has imposed obvious restrictions on their detailed observation. Several can be seen with the unaided eye, but only as misty patches. One of them is M44, generally known as Praesepe or the Beehive. It was well known in ancient times, but its real nature did not become apparent until after the invention of the telescope. It lies in Cancer at a distance of about 575 light-years. The characteristics of its stars are similar to those of the Hyades. The brightest are white giants and orange giants, while the majority form a well-defined dwarf sequence that extends down to at least type K.

Another conspicuous cluster, also known since antiquity, is h and Chi Persei. It lies in the sword-handle of Perseus, almost midstream in the Milky Way, at a distance of about 7,340 light-years. Actually, as the name suggests, it consists of two clusters, but it is generally

NGC 4755, the "Jewel Box" of the southern Milky Way. A cluster of stars near the Southern Cross, faintly visible to the unaided eye.
—MOUNT STROMLO OBSERVATORY

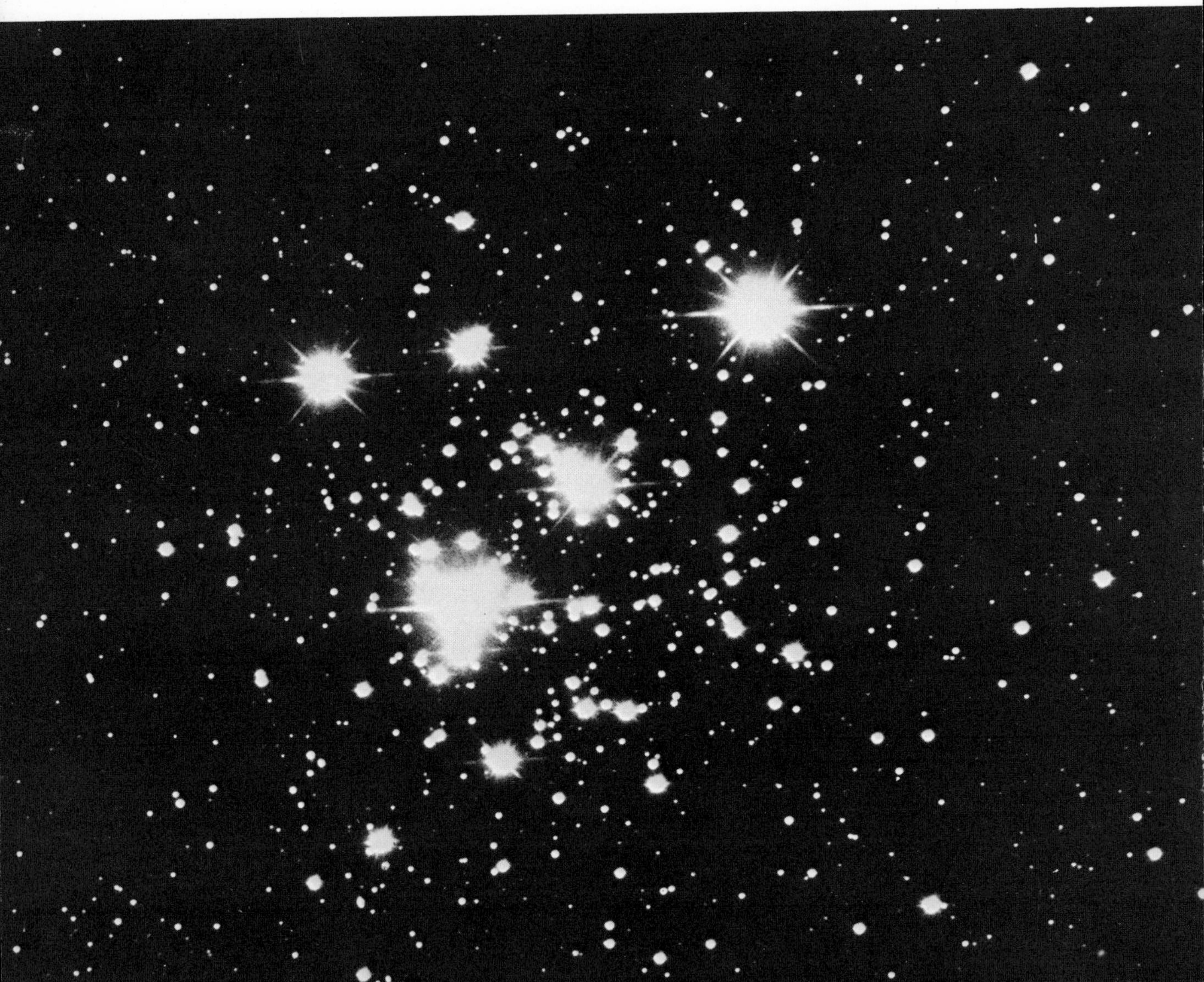

treated as a single object with the name Double Cluster. Once again main sequence stars are numerous, but the object owes its striking telescopic appearance to a liberal sprinkling of supergiants. Some of these, of spectral type O and about 25,000 times more luminous than the sun, must be relatively young. The others, red supergiants, represent stars of similar age which have forged slightly ahead in their evolutionary development.

A similar cluster in southern skies is NGC 4755, beside the Southern Cross. John Herschel aptly called it The Jewel Box since its stars have highly contrasting colors. Its distance is about 7,700 light-years and its "jewels" are three whitish supergiants and a distinctly red supergiant. The cluster certainly contains at least two hundred stars but the membership of the red supergiant, Kappa Crucis, has not yet been established.

M67 IN CANCER Finally, and in complete contrast to the Double Cluster and NGC 4755, is M67, a faint telescopic open cluster near Alpha Cancri. The general faintness of its three hundred or more stars is due partly to their distance, 2,700 light-years, and also to their comparatively low luminosities. The brightest members lie between absolute magnitudes +2 and 0 and are therefore roughly 15 to 100 times as luminous as the sun. They are subgiants rather than giants, but they do at least form a definite and continuous spectral sequence. Practically all the cluster stars are below spectral type A. If the cluster ever had A, B, and O stars, they have long since passed through their evolutionary development and are now white dwarfs. None of the latter has been detected, but this is not surprising in view of the cluster's great distance.

Evolutionary Trends in Star Clusters

The individual peculiarities of these and other clusters become apparent when the member stars are plotted on an H-R diagram. In the composite picture the brighter stars of each cluster form a giant sequence which branches away from the main sequence. Assuming that the members of a cluster had a common origin in space and time, the series of turnoff points from the main sequence is analogous to the stratification of fossils in the ground, the age increasing with descent. Each branch, moreover, can be regarded as an evolutionary track for stars of a particular mass. This assumes that

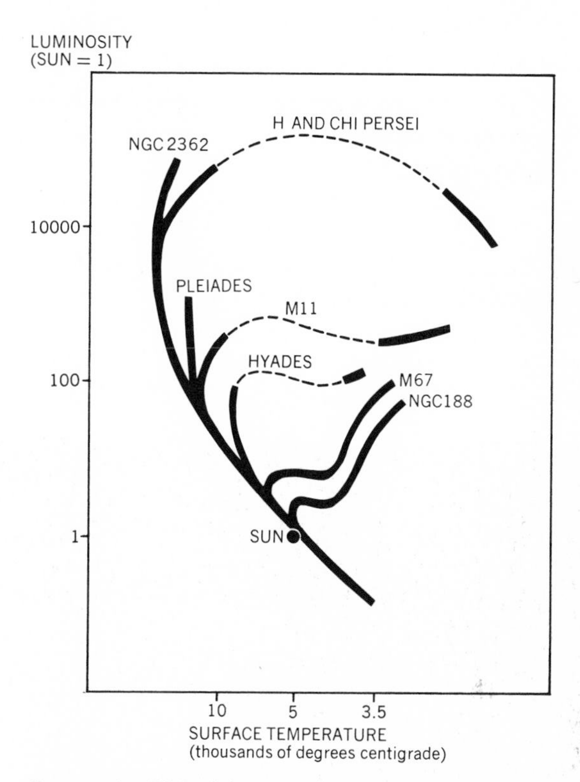

Composite H-R diagram for several galactic clusters. Each giant branch can be regarded as the evolutionary track of stars of a particular mass. The higher the point of turnoff or branching the younger the cluster. In old clusters like NGC 188 and M67 stars slightly more massive than the sun have already evolved well away from the original main sequence.

the giants in any one cluster are similar in mass, but this is bound to be so since they spend so much more of their time on the main sequence than on the giant branch. A striking feature is the gaps in the giant branches; the reader might well wonder why there are no F-, G-, and K-type giants in the Double Cluster, no F-type giants in the Hyades, and so on. The usual explanation is that stars evolve rapidly away from the main sequence and do not slow down until they

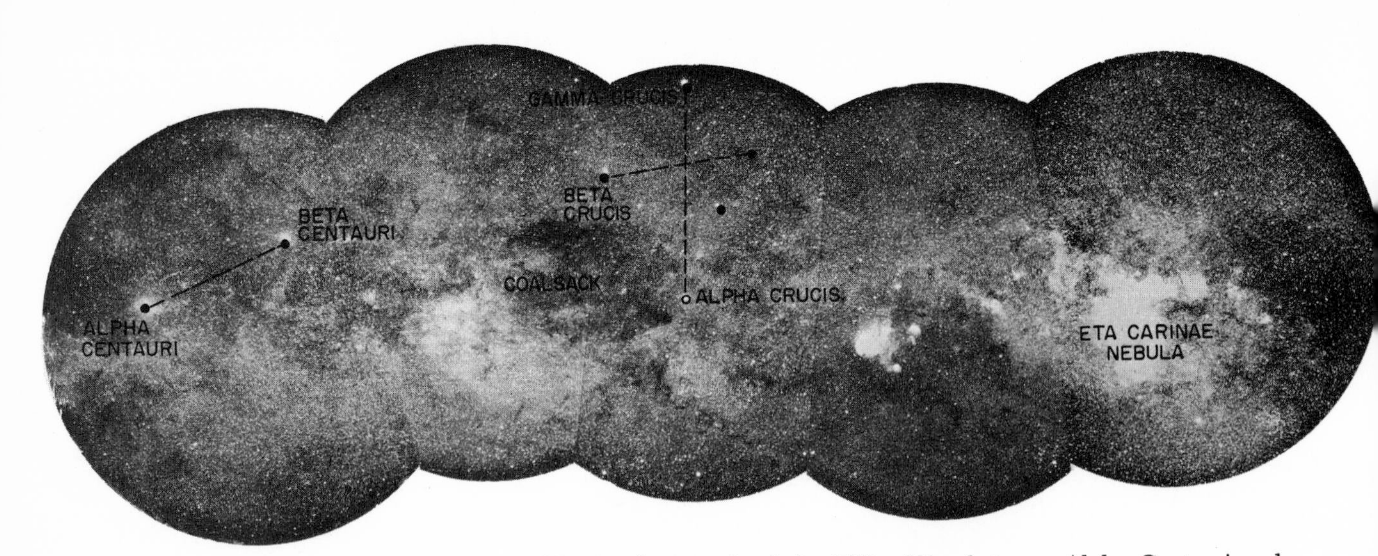

Mosaic photograph of the Milky Way between Alpha Centauri and the Eta Carinae nebula, showing the Coalsack and the Southern Cross.
—MOUNT STROMLO OBSERVATORY

have traversed the gap and are near to becoming full-blown giants.

H-R diagrams of open star clusters show that the Pleiades is by no means the youngest group. The Double Cluster and NGC 2362, among others, are considerably younger, with estimated ages of just a few million years. The Hyades and Praesepe, with lower main-sequence turnoffs, are of intermediate age, while M67 and NGC 188 are among the oldest known. According to Sandage the latter is over 20,000 light-years away and about 14 to 16 billion years old.

The curious kink in its subgiant branch (and also in the branch of M67) illustrates a feature predicted by Hoyle from theoretical considerations. When a star of moderate mass leaves the main sequence it first becomes slightly hotter and therefore bluer. The surface temperature then progressively decreases and it grows gradually redder.

Well over 550 objects are now recognized as galactic, or open, clusters. They contain anything from 20 to perhaps 2,000 stars and lie within a distance of about 16,000 light-years from the sun. This is only a small part of the galaxy so there must be many more beyond the reach of present-day telescopes. The known ones lie in or near the Milky Way and so are all subject, although by different amounts, to the reddening and dimming effects of the interstellar medium.

Interstellar Absorption

The first conclusive proof of a galactic absorption more widespread than that localized in dark nebulae was provided by R. J. Trumpler of Lick Observatory. In the late 1920's Trumpler estimated the distances and dimensions of about a hundred open star clusters. To his surprise their calculated dimensions increased with distance, a most improbable result. Interstellar space was not transparent as had been generally supposed. An extremely rarefied galactic haze both absorbed and scattered starlight and made distant clusters appear fainter and redder than they really were. His distance estimates were excessive, and their correction proved no easy task. The best he could do at the time was to suppose that the clusters were roughly equal in actual size and that the haze was of uniform density throughout. He then found that the haze extended to about 300 to 450 light-years on either side of the general plane of the galaxy and had a coefficient of photographic absorption of 0.67 magnitude per 3,000 light-years.

Later work in this field showed that both assumptions took too much for granted. For one thing the galactic haze is not uniform but has innumerable irregularities in concentration. In the sun's neighborhood alone the coefficient of photographic absorption can be anything between 0.4 magnitude and 3 magnitudes per 3,000 light-years. This brings about corresponding difficulties and uncertainties in determining the absolute magnitudes of stars in fairly near clusters like the Hyades and Pleiades, let alone those in much more remote clusters.

A convenient way of investigating interstellar absorption is to determine the color index of a star from its spectral lines and then compare this with the apparent color index. Stars of spectral types O and B are well suited for this purpose since they have high luminosities and can be seen over great distances. It so happens that globular clusters contain many highly luminous giant stars. Because they also lie at great distances from us and are not restricted to the region of the Milky Way, they are excellent landmarks for testing the extent and effect of the galactic haze, as well as determining the overall extent of the galaxy.

Globular Clusters

To the unaided eye globular clusters are quite insignificant objects. The brightest, Omega Centauri—NGC 5138—lies in the Southern Hemisphere and looks like a hazy star of about the fourth magnitude. In reality it is an immense ball-like swarm of at least several hundred thousand stars. John Herschel described it as "a noble globular cluster, beyond all comparison the richest and largest object of the kind in the heavens." Like other globulars its stars are concentrated toward its center, but if all the stragglers on the outside are included its overall apparent diameter is slightly larger than that of the full moon. Since its distance is 22,000 light-years, the linear diameter must be of the order of 200 light-years.

Omega Centauri Omega Centauri has a slightly elliptical outline, so its actual shape is that of a prolate spheroid. This is now known to be the result of the cluster's rotation. G. A. Harding of the Radcliffe Observatory, Pretoria, has recently determined the radial velocities of several stars in the cluster and found differences that indicate that the cluster makes one complete rotation in a little less than 10 million years. Many other clusters have similar elliptical outlines, so rotation may well be a property of them all.

47 Toucani Another fine southern object is 47 Toucani, near the Small Clouds of Magellan. Its distance is about the same as that of Omega Centauri, and its brightest stars are as faint as magnitude 13. They are very densely packed toward the center and give the cluster an extremely compact but elliptical appearance. The cluster can be seen quite easily on a clear dark night and has often been mistaken for a comet.

M13 in Hercules When William Herschel wrote that globular clusters were "the most magnificent objects that can be seen in the heavens," he had particularly in mind M13, a naked-eye globular between Eta and Zeta Herculis, the brightest of its kind in northern skies. "This is but a little patch" wrote its discoverer Halley in 1716, "but it shows itself to the naked eye when the sky is serene and the moon absent." Messier, using a small telescope, described it as a "nebula without stars," but Herschel, with much larger instruments, resolved it into stars. In his estimation some fourteen thousand stars were "cribb'd, cabined, and confined" in Halley's "little patch." According to modern estimates its distance is about 34,000 light-years.

The 118 known globular clusters have a great range in size and stellar content. The stars of some are much more densely packed toward the centers than those of others; some have a few thousand stars while others have several hundred thousand or, maybe, over a million. Their full angular extents are difficult to determine, but it seems that their linear diameters can be anything from about 50 to 200 light-years. Most photographs of them are the result of fairly long time-exposures since these bring out the fainter stars. This means that their centers are over exposed and look like solid masses of stars. Actually their stars are probably separated from one another by several million million miles, even in the most crowded parts.

Most globular clusters are so far away that only their brightest stars can be distinguished. Of these, the most luminous are red giants, followed by giants and subgiants of spectral types A, F, G, and K. Still farther down the scale are main-sequence stars, but those of spectral types K and M, if they exist at all, are too faint to be seen.

The H-R diagram of a typical globular cluster differs markedly from one based on stars in

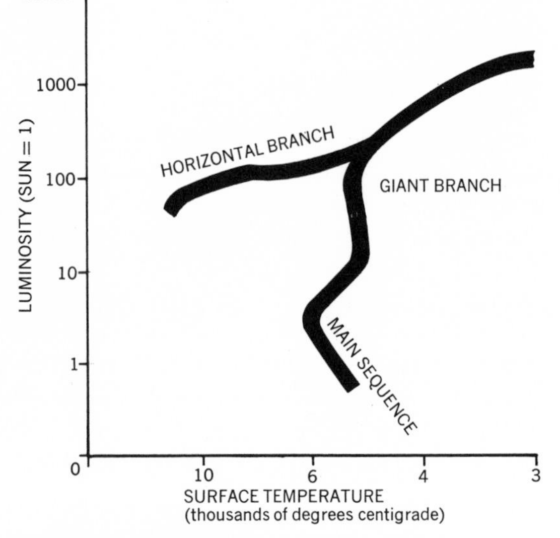

Simplified H-R diagram for M3, a globular cluster in Canes Venatici.

the sun's neighborhood, or again, on the members of a galactic cluster like the Pleiades. This is not because of any inadequacies in the available data but rather because globular clusters consist mainly of old stars. Many of their members have already left the main sequence and are now passing through the giant and white dwarf phases. Further evidence of old age is the fact that they contain hardly any O- and B-type stars and little or none of the interstellar gas and dust associated with young stars. But they do contain RR Lyrae stars, Population II Cepheids, and several hot subdwarfs, and these are all objects in a fairly advanced stage of their evolution.

The position of the turnoff point, or the point on the main sequence where the giant branch originates, in the H-R diagram of a typical globular cluster indicates that its age is of the order of 10 billion years. Yet oddly enough many globulars have an appreciable number of main-sequence stars *above* the turnoff point. These stars must be comparative newcomers, for in 10 billion years any original hot blue stars would have evolved well away from the main sequence. We can only suppose that the cluster stars, like those of several galactic clusters, were not all formed at the same time.

Some support for this idea is found in the dark streaks seen on some globular clusters. They are particularly noticeable on M13, M3, and M2. Their existence has been known for nearly a century, but until recently, they were generally regarded as effects produced by chance groupings of stars. It now seems more likely that they are patches of dark nebulosity, but whether they are in the foreground or actually in the clusters is debatable. If the latter is true they could be dark globules or protostars. Perhaps they represent the forerunners of a second generation of stars in the sense that they are formed from material shed by the first generation. But this is all highly speculative. The only definite evidence of diffuse matter in globular clusters is the planetary nebula in M15.

CHAPTER XI

THE GALAXY OR MILKY WAY SYSTEM

We owe the first detailed description of the Milky Way to Ptolemy. "The Milky Way," he wrote, "is not a circle, but a zone, which is almost everywhere as white as milk, and this has given it the name it bears. Now, this zone is neither equal nor regular everywhere, but varies as much in whiteness as in shade of color, as well as in the number of stars in its parts, and by the diversity of its positions; and also because that in some places it is divided into two branches, as is easy to see if we examine it with a little attention." He then described its course, pointing out where it appeared "rather dull" (in parts of Argo, the Ship), and where "dense" and "like a smoke" (in Sagittarius, the Archer).

The center line of the Milky Way passes within about 30 degrees of the north celestial pole and runs through Cassiopeia, Perseus, and Auriga. It then passes between Orion and Gemini, through Monoceros to Argo, Grus, and Centaurus. Here it divides into two streams, the brighter of which passes through Ara, Scorpius, Sagittarius, and Aquila and rejoins the other in Cygnus. Its width varies considerably, being about 45 degrees wide between Orion and Canis Minor, and only about 4 degrees in some other places. The brightest parts are in Sagittarius and Scorpius, but it is also prominent in Aquila, Cygnus, and Centaurus.

Galactic Coordinate System

The Milky Way divides the celestial sphere into two roughly equal parts, forming a celestial equator of sorts, but its irregular outline and width prevent it being used for this purpose. We can, however, imagine a great circle drawn as nearly as possible through the center of its course. The plane of this circle, known as the galactic plane, can then be used to define the position of a star or nebula in terms of galactic latitude and longitude. The galactic circle is inclined at about 63 degrees to the celestial equator and intersects it in Aquila and Monoceros. Prior to 1958 the intersection in Aquila served as galactic longitude 0 degrees, but since then the latter has been referred to the direction of the galactic center, in Sagittarius. The northern pole of the galactic circle (galactic latitude +90 degrees) is situated in Coma Berenices, and the southern pole (galactic latitude −90 degrees) is in Sculptor. Galactic coordinates are very important in investigations of the motions and distributions of stellar objects, for the galactic circle is in a very real sense the "ecliptic of the stars."

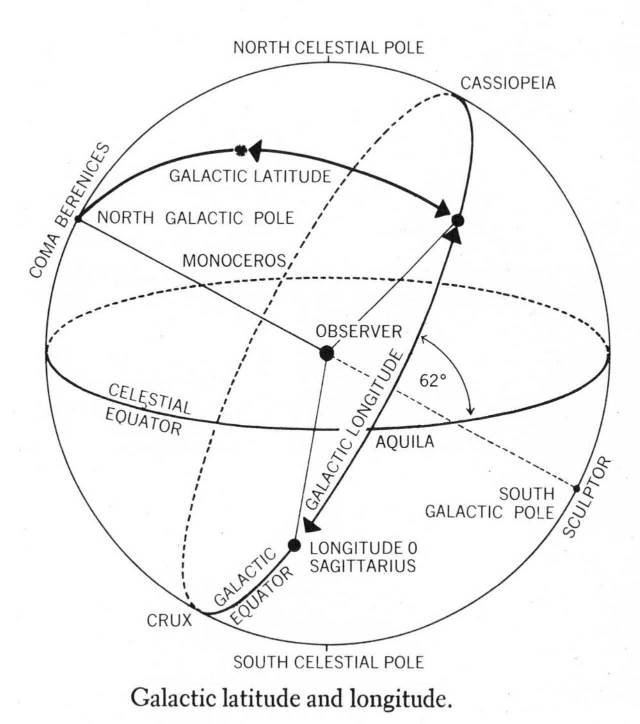

Galactic latitude and longitude.

Disk Theory of the Milky Way

The "disk-theory" of the Milky Way was first proposed in 1750 by T. Wright of Durham.

Wright published a series of lectures in which he suggested that the Milky Way was not a vast collection of faint stars but an optical effect brought about by the central position of the sun in a flattened stellar system. The German philosopher Kant supported Wright's theory, but claimed the stellar system centered on Sirius and owed its flattened form to its rotation. He also suggested that there was an infinite number of other "Milky Ways," star systems similar to ours, but so far away as to look like round- or oval-shaped nebulae. Neither worked out the consequences of the hypothesis.

It was left to William Herschel to put these ideas to the test of observation and to arrive at the conclusions outlined in chapter I. The very fact that faint (and therefore distant) stars showed a decided preference for the Milky Way implied that they formed a flattened system. But Herschel had to admit later that his assumption that the stars were distributed evenly in space failed to meet the facts. "The immense starry aggregation of the Milky Way," he wrote in 1802, "is by no means uniform. The stars of which it is composed are very unequally scattered, and show evident marks of clustering together into many separate allotments."

GALACTIC ABSORPTION What Herschel and his immediate successors did not (and could not) allow for were the effects of galactic absorption. They thought that the extent to which they could probe was limited only by the powers of their telescopes, but quite often the limit was set by the distribution of absorbing material. We now realize that our position for surveying the galaxy is not a particularly good one. In the first place the solar system lies near the plane of the galaxy: this follows from the fact that the Milky Way divides the celestial sphere into two roughly equal parts. (The northern part is slightly larger than the southern so the sun lies to the north of the plane). In the second place the absorbing medium, like the stars of the galaxy, has a general concentration towards the galactic plane. So when we look toward the center of the galaxy; or in the direction of Sagittarius, we cannot see its innermost parts. The region of Sagittarius is the brightest part of the Milky Way—in some places the star density is so high that star images on photographs merge into one another to form a continuous background. In addition, dark lanes break up the rich field of stars into great cloudlike masses, appropriately called "star clouds."

Galactic absorption acts all along the Milky Way. Gaps in the dark clouds enable us to detect more distant star clouds, globular clus-

Mosaic of the Milky Way, from Sagittarius to Cassiopeia. Note the distinct belt of obscuration, much as in an edge-on view of a spiral galaxy.

—MOUNT WILSON AND PALOMAR OBSERVATORIES

ters, and galaxies, but these are all veiled to some extent by the underlying galactic haze. The latter becomes less effective in high galactic latitudes, and least of all at the galactic poles. For this reason the very distant galaxies are comparatively scarce in low galactic latitudes but extremely numerous in the regions of the galactic poles.

Star Streaming By the end of the last century, investigations of proper motions in bulk had revealed only one general star drift—the reflection of the sun's motion. Yet no two authorities could agree on the precise direction of that motion. The proper motions of different groups of stars merely showed that the solar apex, or point on the celestial sphere toward which the sun is traveling, lay somewhere in the region of Lyra and Hercules.

The determination of the sun's speed proved even more difficult. It depended on estimates of the radial velocities of individual stars, and the value obtained was always relative to the group of stars considered. The reference stars used in these studies had their own individual space motions, but it was thought that these were more or less random. This encouraged the expectation that the proper motions and radial velocities of the stars in a particular area of the sky would tend to cancel one another, leaving a reliable value of each component to determine the true effect of the sun's motion.

In 1904 Kapteyn put this expectation to the test. Breaking with tradition he allowed for the effect of the sun's motion on the proper motions of several bright stars and found that they were not random. Instead, they showed that the brighter and nearer stars belonged to one or other of two interpenetrating stellar swarms or streams. The discovery came as a complete surprise, but was brilliantly confirmed and developed by Eddington in the years that followed. Boss had produced a catalog of the proper motions of six thousand naked-eye stars, so Eddington had plenty of material to work on. He found, among other things, that when the sun's motion was allowed for, the two streams appeared to move in opposite directions, almost exactly in the galactic plane.

Further investigations revealed that various groups of stars had similar stream motions, but Kapteyn's two streams still remained very much in evidence. The fact that they moved in the galactic plane suggested that they were due to some fundamental property of the galaxy as a whole. Kapteyn thought that some stars revolved about the center of the galaxy in one direction while others revolved in an opposite direction. Eddington suspected that the galaxy had two spiral arms and that stars were moving along these arms in contrary directions.

Shapley Estimates the Size of the Galaxy In 1918 Shapley announced the results of his studies of the space distribution of globular clusters. Astronomers were surprised to hear that these objects were about equally distributed on each side of the galactic plane and formed a large flattened system fully 250,000 light-years in diameter by about half that amount in thickness. The center of the system lay in the direction of Sagittarius at a distance of about 65,000 light-years from the sun. Shapley had moved the center of the galaxy from the sun to the center of the system of globular clusters and given the galaxy an overall diameter of about 300,000 light-years. Kapteyn and his pupil P. J. van Rhijn thought that Shapley had placed too much reliance on the Cepheid period-luminosity relationship and that his estimated distances of the globular clusters were unreliable. In their opinion the distribution of faint stars as revealed by photographic star counts pointed to a much smaller system whose center lay fairly close to the sun.

The Rotation of the Galaxy

The problem was resolved in 1926–1927 by Lindblad and Oort. They showed that Kapteyn's two star streams could be accounted for if the galaxy rotated about a center in the direction of Sagittarius. The rotation presumably accounted for the galaxy's flattened form, but the galaxy did not rotate all in one piece like a solid body. Otherwise stars in the sun's neighborhood would have individual movements of their own but show no systematic trends. Instead, the orbital motions of the stars resembled those of the planets. Stars at a great distance from the center of motion traveled more slowly than those closer to the center. Those farther away from the center than the sun appeared to lag behind the sun, while those nearer the center appeared to forge ahead. These differential motions affected both the radial velocities and proper motions of nearby stars and explained why they appeared to drift in two opposite directions in the galactic plane.

Further information about the rotation of the galaxy was provided in the 1930's by J. S. Plaskett and J. A. Pearce of the Dominion Astrophysical Observatory, Victoria, B. C. Their observations of the proper motions and radial velocities of O- and B-type stars revealed the differential effects required by theory and at the same time confirmed Oort's basic idea that the mass of the galaxy was strongly concentrated towards the center. They also followed up an earlier discovery (made by G. Hartmann in 1904) that the spectra of distant spectroscopic binaries contained "stationary" absorption lines due to ionized calcium. They found that this gas, distributed throughout interstellar space, shared in the galactic rotation and had an almost uniform distribution. The rotation was a consequence of the gravitational field of the galaxy and not an effect peculiar to the stars.

THE SUN'S ORBITAL VELOCITY Another conspicuously early achievement was the determination of the sun's orbital velocity. Relative to nearby stars the sun had a velocity of about 12 miles a second, but relative to some of the galaxies its velocity was found to be about 190 miles a second. This was only a rough estimate, however, for the galaxies were known to be in motion relative to the center of the galaxy. An alternate value was obtained by measuring the radial velocities of twenty-six globular clusters. It came out at about 170 miles a second but rested on the questionable assumption that the clusters formed a nonrotating system.

Distance of the Sun from the Center of the Galaxy

Once astronomers had decided on a value for the sun's orbital speed, they could deduce its distance from the center of the galaxy. They had to suppose that the orbit was more or less circular and that the mass of the whole system was concentrated toward its center, but these assumptions were thought to be quite reasonable. The very fact that the region in the sun's neighborhood did not rotate like a solid body meant that the mass had a nonuniform distribution. For many years the sun's speed relative to the center was taken as 170 miles a second, so the corresponding radius of its orbit came out at 30,000 light-years, or roughly one half of Shapley's original estimate. The period of the sun's revolution was then 200 million years and the galaxy had a mass of the order of 100 billion suns.

Astronomers later favored a smaller value of 27,000 light-years for the radius of the sun's orbit. This was obtained by Oort and W. Baade working quite independently. Oort's result was based on the radial velocities of clouds of neutral hydrogen, and Baade's on observations of RR Lyrae stars in the field of the globular cluster NGC 6522. These stars lie in the central bulge, or nucleus, of the galaxy and are reasonably although not entirely free from the effects of interstellar absorption. If we adopt the Oort-Baade distance, the corresponding value for the speed of the sun is 135 miles a second. More recent radio observations, however, have led investigators to turn to the higher values of 33,000 light-years for the radius and 160 miles a second for the sun's speed.

According to theory, objects that stay in or near the galactic plane move in almost circular orbits* with comparatively large velocities. Dark and bright nebulae, open star clusters, hot O- and B-type stars, and the great majority of normal main-sequence stars have just these characteristics, but their velocities relative to the sun, itself traveling at great speed, are quite small (up to about 20 miles a second). Objects traveling in orbits whose planes are steeply inclined to the galactic plane have comparatively small velocities in highly elliptical orbits. They

* The idea that these and other stellar orbits in the galaxy are closed curves is only a convenient fiction. Since the mass of the galaxy is not concentrated at a point, the orbits are complicated loops.

include globular clusters, RR Lyrae stars, and Population II Cepheids, all of which appear to travel in a direction opposite to that of the sun with large velocities (up to about 250 miles a second).

Stellar Populations

These distinctions form the basis of the modern concept of two stellar populations. Objects whose orbits are restricted to the plane of the galaxy are said to belong to Population I, while those whose orbits carry them well away from the plane belong to Population II.

Population I objects are generally younger than those of Population II. Hot O and B stars, both as individuals and in groups, are invariably associated with the interstellar gas and dust in which they had their birth. Open star clusters like the Double Cluster and the Pleiades have remained intact despite the disrupting influence of external forces. These and other Population I objects form the galactic "disk," a comparatively thin layer about 4,000 light-years deep, or 1/25 of the overall diameter of the galaxy. Red giants and red supergiants, on the other hand, are well advanced in age, so also are the metal-poor RR Lyrae stars and hot subdwarfs. The same can be said of globular clusters: they appear to be fairly free from gas and dust, so only a few new stars, if any, can be formed in them. These and other Population II objects form the galactic "halo," a structureless and roughly spherical system some 150,000 light-years in diameter.

We can think of the galaxy as a combination of two interpenetrating and concentric systems. That of Population II, heavily concentrated towards the center, forms a kind of substratum in which the Population I system is embedded. The heavy central concentration is the nucleus, flattened in form owing to rapid rotation and with a polar diameter of about 16,000 light-years. If the galaxy could be stripped of its spiral arms, and therefore of Population I, it would resemble a gigantic globular cluster with numerous much smaller globular clusters embedded in it. Most of the stars in the galaxy belong to this "cluster" and therefore account for the greater part of the light emitted by the galaxy as a whole.

Since the two populations interpenetrate, we can expect to find in the solar neighborhood visiting Population II stars among the more permanent residents of Population I. The metal-poor subdwarfs are in this category. Because these stars travel with high velocities relative to the sun in a direction opposite to that of the sun's orbital motion, they belong to Population II. Again, most globular clusters are old enough to have passed through the disk several times. Several have fairly low galactic latitudes and one, NGC 6522, lies near the galactic center. Similarly a member of Population I may appear in the region of Population II. The open star cluster Coma Berenices, for example, lies near the north galactic pole and therefore in a high galactic latitude, but it is really only 260 light-years "north" of the sun and therefore well inside the disk.

Spiral Structure of the Galaxy

We now know that the galaxy has a spiral structure. The distinct spiral patterns of galaxies like M51, the Whirlpool Nebula in Canes Venatici, encourages this view but does not prove it. Many galaxies have only a faint semblance of spiral structure and some have none at all. In 1951, however, W. W. Morgan announced that he and his colleagues at the Yerkes Observatory had been able to trace sections of three spiral arms. They found that certain groups of O- and B-type giants and supergiants fell along three fairly well-defined and almost parallel bands. The first, named the Orion Arm, contains the sun and stretches for about 12,000 light-years from Cygnus to Monoceros. The second, or Perseus Arm, contains the Double Cluster and is some 5,000 light-years farther away from the galactic center. The third, or Sagittarius Arm, is closer to the galactic center than the sun, but only part of it can be observed from northern observatories.

By a curious coincidence 1951 also saw the introduction of a powerful new technique which soon led to a much more comprehensive picture of the spiral structure of the galaxy. Radio astronomers in the United States, Holland, and Australia successfully observed the 21-centimeter radiation of neutral atomic hydrogen. Its existence had already been predicted ten years earlier by the Dutch astronomer H. C. van de Hulst, but its detection was delayed owing to World War II.

The great practical importance of this radiation lies in its ability to pass through interstellar dust clouds without hindrance. Observations at 21 centimeters enable us to study the distribution and motions of neutral hydrogen in regions far outside the optical range and therefore beyond the center of the galaxy. They show that the hydrogen is largely confined to the galactic disk and to the members of Population I. It is also very unevenly distributed, being concentrated in clouds and sheets, some of which are of enormous size. The clouds have their own individual motions but generally share in the rotation of the galaxy. Most important of all, they are concentrated to the spiral arms and so enable us to map the spiral structure of almost the entire system.

M51, the Whirlpool, a spiral galaxy in Canes Venatici similar to our own galaxy. The bright blob linked to one of the spiral arms is the companion galaxy NGC 5195.
—MOUNT WILSON AND PALOMAR OBSERVATORIES

Most of the mapping has been done by radio astronomers at Leiden and Sydney. They have also made a great number of radial-velocity measurements (based on Doppler shifts of the 21-centimeter line) that show that the rotation of the galaxy is more complex than was at first thought. The velocity of rotation increases from about 120 miles a second at a distance of 8,000 light-years from the galactic center to a maximum of 170 miles a second at a distance of 20,000 light-years. It then gradually decreases to reach 140 miles a second at the distance of the sun and presumably drops to even lower values in the outermost regions.

THE 3-KILOPARSEC ARM Some years ago the Leiden school discovered the "3-kiloparsec arm," a complex armlike structure which passes between the sun and the galactic center. A similar structure has since been located on the other side of the galaxy at the same distance from the center. The most remarkable property of these structures is their outward motion of 40 or more miles a second. They are therefore thought to be parts of a rotating and expanding shell of gas which presumably had its origin in the nucleus. Armlike structures further away from the center exhibit similar deviations from circular motion and have led F. J. Kerr at Sydney to conclude that the outward flowing gas passes the sun at a velocity of about 5 miles a second.

It is tempting to think of the 3-kiloparsec arm as part of an embryo spiral arm which will eventually take the place of one of the present spiral arms when its supply of hydrogen has become exhausted. Also, that it is an aftereffect of a giant explosion which occurred in the central regions some 50 million years ago. Similar outward motions have been detected in M31 and M51, while other galaxies are subject to much more violent and extensive outbursts.

THE SPIRAL ARMS The galaxy appears to

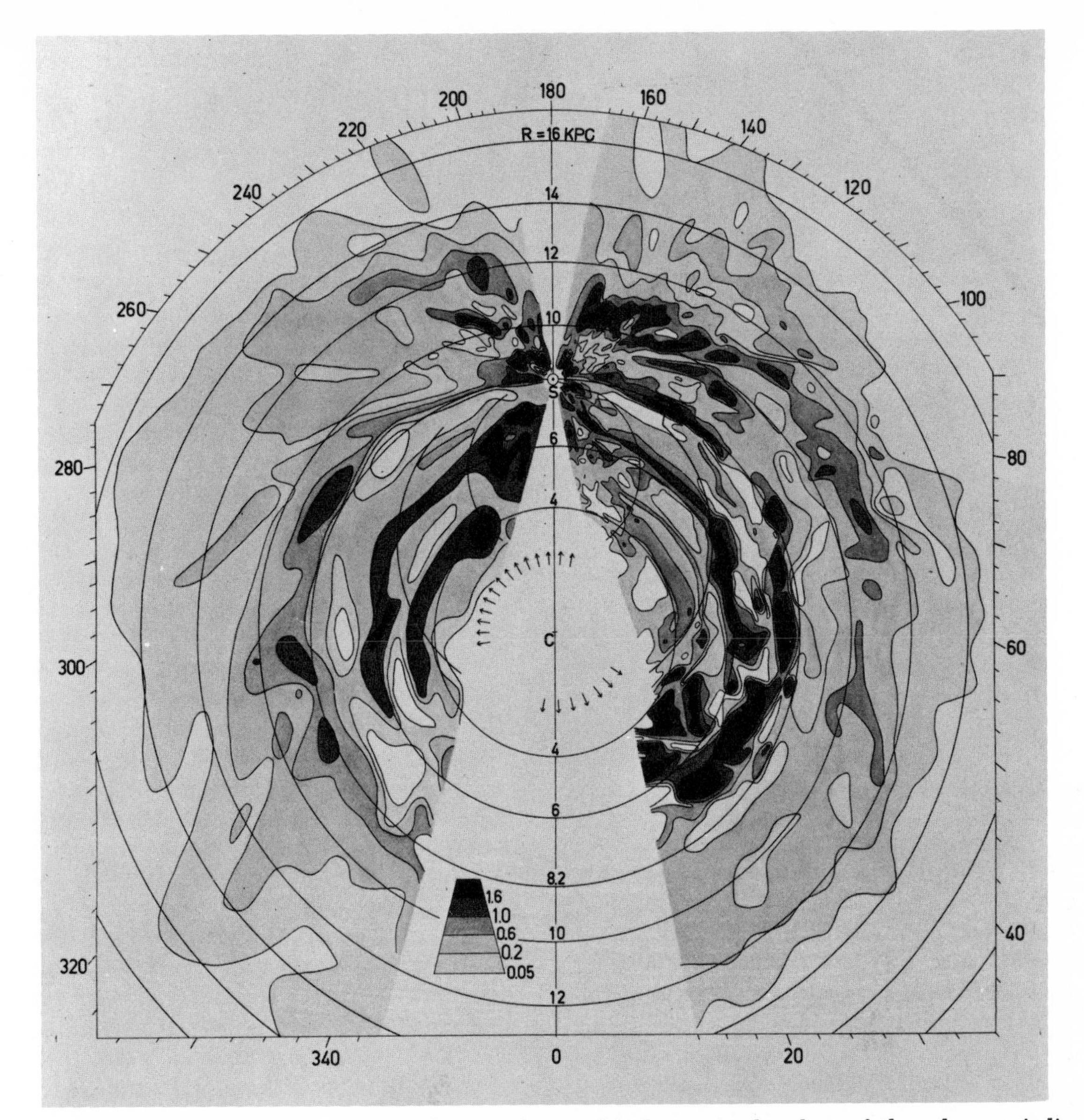

Distribution of neutral hydrogen in the plane of the galaxy as indicated by 21-centimeter radio observations made at the Observatory of Leiden, the Netherlands, and at the Radiophysics Laboratory, CSIRO, Sydney, Australia. The different shadings show the density of hydrogen atoms per cubic centimeter. The center of the galaxy is at C, the sun at S. Distances from the center are indicated in kiloparsecs (1 kiloparsec = 3262 light-years). Numbers on the periphery are galactic longitudes as seen from the sun. The two empty sectors are regions where hydrogen distribution cannot be unraveled from radio data. —J. H. Oort, Leiden Observatory

have two almost circular and roughly continuous spiral arms and to be fairly typical of a "semi-open" spiral. The arms curve away gently from the nucleus and after making two or three turns trail off into intergalactic space. Bok, working under southern skies, found that the sun and the Coalsack were located on the inner border of what he called the Carina-Cygnus Arm. He considers that this band, along with its two short branches or spurs (the Orion Spur and Vela Spur), is the major local spiral feature. W. Becker, however, sees the Carina-Cygnus

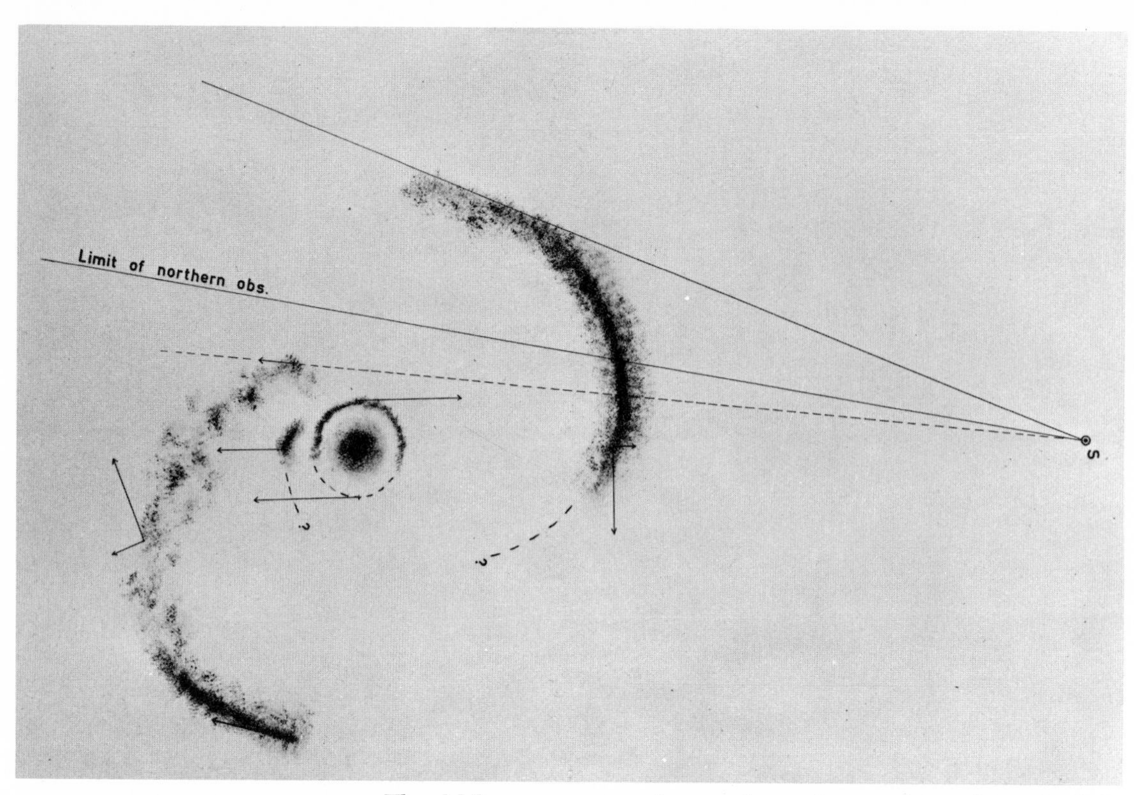

The 3-kiloparsec arm and central rotating patch as determined by observations of the 21-centimeter line of neutral hydrogen.
—J. H. Oort, Leiden Observatory

Arm as part of the Sagittarius Arm, with the Orion Arm going out of sight in the direction of Vela. It is now fairly well established that one spiral arm is represented by the Perseus Arm, and one turn or convolution later, by the Sagittarius Arm.

One of the many theoretical problems in this field of study is how to explain the relative permanence of the spiral arms. Fundamentally they are patterns of gas, for the stars born in them tend gradually to wander into other regions. The highly luminous blue giants and supergiants scattered along them are young and comparatively short-lived stars. These giants can radiate at a high rate for only 1 million to 10 million years, whereas the galaxy itself is thought to be at least 10 thousand million years old. They may, of course, have arisen from the debris of an earlier generation of stars, but it seems likely that the original hydrogen content of the arms would otherwise now be almost depleted. Perhaps the hydrogen is replenished by the flow from the center of the galaxy, but if so how is the flow maintained and at what rate is the central supply itself being depleted? The mystery remains to be solved.

Optical and radio observations indicate that the galaxy is a "winding" as distinct from an "unwinding" spiral. How, then, can the arms preserve their shape? The pressure of a large outflow of gas would probably prevent them from winding themselves into a knot after a few revolutions, but the observed flow is much too small to be effective in this respect. One suggestion is that they have invisible backbones in the form of magnetic lines of force, but recent estimates put the general field strength of the galaxy at only 5 millionths of a gauss at most. Stronger fields are associated with some interstellar clouds but they are too localized to play a significant part in controlling the shape of the spiral arms.

The Magnetic Field of the Galaxy

The observational basis for an interstellar magnetic field was first provided by W. A. Hiltner

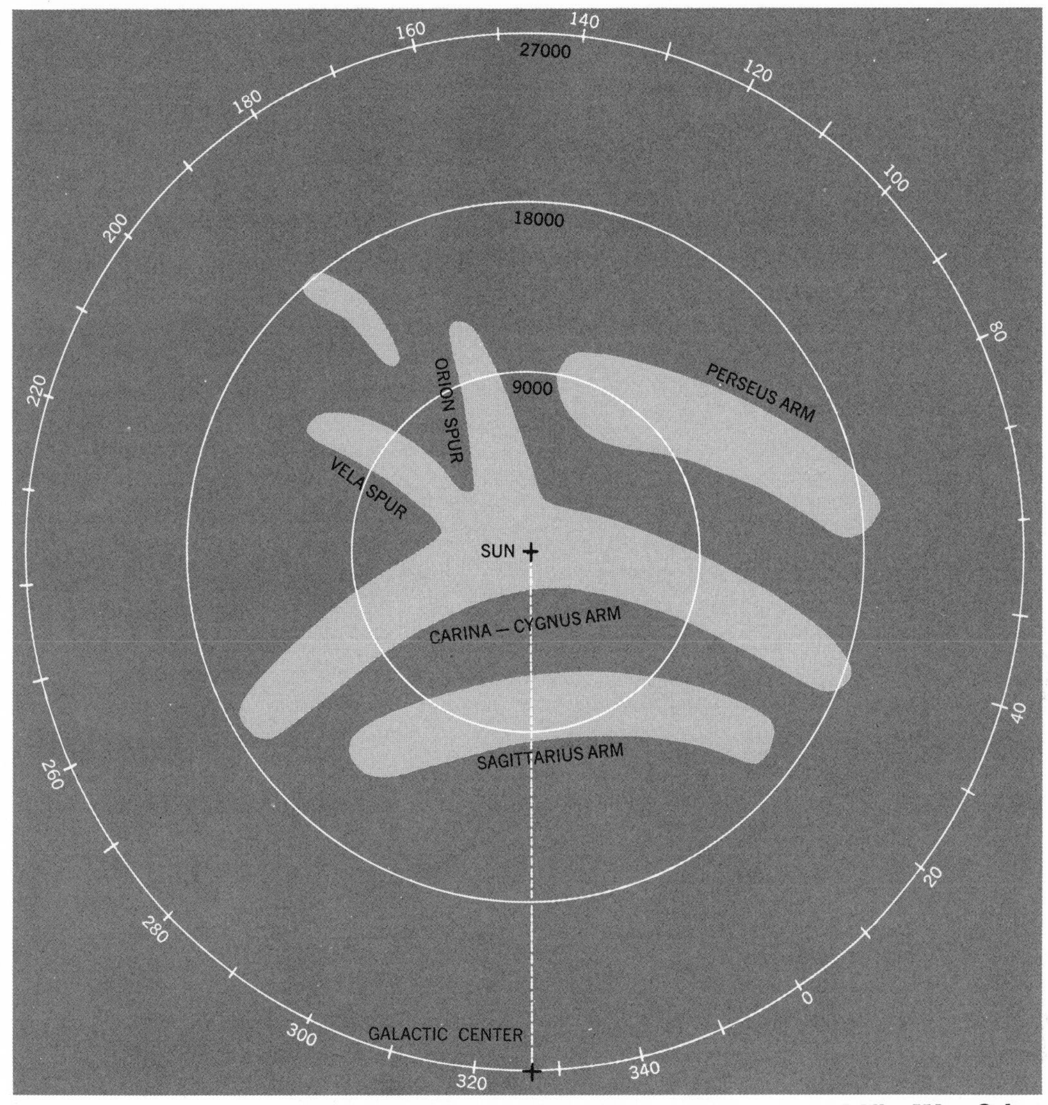

B. J. Bok's concept of the spiral structure of the Milky Way Galaxy within about 18,000 light-years of the sun. (Based on a diagram in "The Arms of the Milky Way" in *Scientific American*, Dec., 1959.)

and J. S. Hall in 1949. Their independent observations showed that the light of some space-reddened stars was plane-polarized. L. Davis, Jr. and Greenstein then showed that this effect could be produced if vast numbers of interstellar grains were aligned parallel to lines of magnetic force.

The grains were imagined to have shapes similar to that of a football and to contain small percentages of iron and other metals. The impacts of molecules in the interstellar gas made them rotate rapidly about their short axes, and in a weak magnetic field they tended to align themselves with their long axes perpendicular to the lines of magnetic force. Davis and Greenstein found that sufficient alignment for polarization could be produced by magnetic field strengths of 10 to 100 millionths of a gauss.

More recently (1962) Hoyle and N. C. Wickramasignhe have proposed that the grains are produced in the atmospheres of carbon stars. As the atmospheres cool the carbon condenses into clouds of grains that are then blown into space by radiation pressure. The grains are graphite particles, possibly coated with ice, and their alignment can be effected by magnetic fields of the order of a millionth of a gauss.

Support for the concept of a galactic magnetic field is provided by continuous radio-frequency radiation. Part of this has its origin in the galactic disk and appears to be associated with the spiral structure. The rest comes from regions well clear of the disk but its intensity is by no means uniform. In both cases the radiation is generated by the synchrotron process, so its presence points to the existence of a large-scale but weak magnetic field in a vast atmosphere or "galactic corona" of extremely tenuous ionized gas.

Synchrotron radiation has a high degree of linear polarization, and by measuring the intensity and direction of this it has recently been possible to outline the general pattern of the magnetic field in the vicinity of the sun. It seems that the magnetic axis of this field almost coincides with the orientation of the spiral arm in which the sun is embedded. More precisely, the magnetic axis lies in the galactic plane and makes an angle of 70 degrees with the direction of the galactic center (longitude 0 degrees). The fact that this angle is somewhat less than 90 degrees is to be expected since the spiral arm is presumably spiraling out from the center.

Early in 1964 D. Morris and G. Berge of the Owens Valley Radio Observatory found that the polarity of some segments of the field is opposite to that of other segments. The reversals occur systematically with galactic longitude and take place in the galactic plane. The direction of the field above the plane is therefore predominantly opposite to that below the plane—a completely unexpected result.

The Nucleus of the Galaxy

It is a great pity that the greater part of the nucleus, or central bulge, of the galaxy lies hidden from view. Most of our knowledge of it is therefore indirect, being based on observations of the nuclei of other spiral galaxies. Red and infrared techniques enable us to probe its outer parts, but so far information about its associated gaseous complex has come from radio studies. The latter have revealed that a strong gaseous ring is being ejected from the nucleus, presumably the result of an explosion about one million years ago. The ring, like the outer 3-kiloparsec ring, is rotating as well as expanding.

Clouds of the Hydroxyl Radical

Another recent discovery is that of small clouds of the oxygen-hydrogen molecule OH, otherwise known as the hydroxyl radical. They give rise to absorption lines in the otherwise continuous spectrum of radio waves due to the synchrotron process. Several have been found near the galactic center, but unlike the rotating rings of hydrogen they appear to be moving directly outwards at speeds of about 60 miles a second.

Curiously enough the OH absorption is well marked near the center but scarcely detectable out in the spiral arms. Its discovery raises a whole crop of new problems, not the least of which is how hydroxyl molecules are formed in or near the nucleus in the first place. At the same time it raises fresh hopes of finding absorption lines due to other molecules, although unfortunately the popular candidate, the hydrogen molecule itself, cannot absorb or emit radiation when its energy state is low.

It is natural to ask whether two stellar populations are found in other spiral galaxies. The answer appears to be "yes," but with an important reservation: no two spirals are precisely alike in size, structure, hydrogen content, and relative abundances of old and young stars. Furthermore, many galaxies have no spiral structure at all and contain only Population II stars. The evolutionary significance of these differences is still subject to considerable speculation, although recent studies have revealed fairly definite trends. There can be no doubt that some galaxies are younger than our galaxy, in development if not in years, and that others are older. But we must first describe these systems in reasonable detail, and in particular, deal with four near neighbors of ours—the two Clouds of Magellan, M31 (the great Spiral Galaxy in Andromeda), and M33 (a spiral galaxy in Triangulum).

CHAPTER XII

NEIGHBORING GALAXIES

As early Portuguese and Dutch navigators sailed farther and farther south of the equator they noticed that there was no bright South Pole star. Instead, two misty patches, about 21 degrees apart and looking like misplaced pieces of the Milky Way, formed a roughly equal-sided triangle with the south celestial pole. The patches were described in 1521 by Antonio Pigafetta after Magellan's voyage round the world and from then on became known variously as the "Clouds of Magellan," "Cape Clouds," and "Nubeculae." But their only value in those days was as guides to the south celestial pole and hence to the direction of the south point and to the latitude of the place of observation.

The Clouds of Magellan

The Large Cloud is mostly in the constellations of Doradus and Mensa: to the unaided eye it covers an area of about 42 square degrees. The Small Cloud lies mainly in Toucana. It is fainter and more uniform in brightness than the Large Cloud and covers an area of about 10 square degrees. Nearby is the globular cluster 47 Toucani, an easy naked-eye object. Otherwise this part of the sky is fairly barren and uninteresting, especially when compared with the richly scattered bright stars of Centaurus, Crux, and Argo over on the other side of the south celestial pole.

The Clouds were first studied in detail by

The Small Cloud of Magellan, photographed in the red light of hydrogen-alpha. The globular cluster to the right of the Cloud is 47 Toucani. —MOUNT STROMLO OBSERVATORY

The Large Cloud of Magellan, photographed in the red light of hydrogen-alpha and showing numerous patches of emission nebulosity.
—MOUNT STROMLO OBSERVATORY

John Herschel, who between 1834 and 1838 surveyed the stars with a reflecting telescope of about 18-inch aperture from a temporary observatory near Cape Town. His main aim was to extend his father's catalogs of double stars and nebulae to include southern objects, for until then southern skies had never been explored systematically with a large telescope. At first he drew the Clouds as he saw them with the unaided eye, noticing that the Large Cloud had a brighter part in the form of a broad bar or axis. He then studied them through the large telescope, but was almost overwhelmed by what he saw.

In the Large Cloud he found clusters of stars, globular clusters, "regular and irregular nebulae properly so called," and large tracts and patches of nebulosity "in every stage of resolution." In those prephotographic days the task of charting so many faint and small objects, all liberally interspersed with stars, required immense skill and patience. Altogether he determined the positions of 641 stars and 278 clusters and nebulae. "I ought to add," he wrote, "that all my own attempts to delineate other than very small portions of the nebula major from the telescope have been completely baffled by the overwhelming complexity of its details. The lesser cloud is less complex—but for that very reason less interesting."

In the Large Cloud he saw S Doradus, later found to be one of the most luminous known stars, and 30 Doradus, also called the Loop Nebula or Tarantula Nebula. He described the latter as "one of the most singular and extraordinary objects which the heavens present" and noticed a small cluster of faint stars near its center. In the Small Cloud he determined the positions of three globular clusters, four clusters of the open type, and well over two hundred stars.

RECENT OBSERVATIONS The next major development came in 1890 with the establishment of a southern branch of Harvard Observatory at Arequipa, Peru. From photographs taken at Arequipa, Mrs. M. Fleming in 1897 detected the variability of S Doradus and Miss Leavitt discovered the first of hundreds of Cepheids that have since been found in both Clouds. Miss Leavitt did not identify the variables in the Small Cloud with Delta Cephei, but when Hertzsprung established the identification in 1914 the stage was set for Shapley's classic work on the size of the galaxy. His announcement in 1917 that the Clouds were more than 60,000 light-years away came as a complete surprise to those astronomers who had in mind a distance of the order of 3,000 light-years.

More recently the Clouds have been studied from an increasing number of southern observatories, but none of these contains a telescope comparable in size to the giant European and American reflectors. At present the two largest telescopes in the southern hemisphere are the 74-inch reflectors at the Radcliffe Observatory, Pretoria, and the Mount Stromlo Observatory, New South Wales, but there are expectations that a 150-inch reflector will be established in Australia in the not too distant future.

DISTANCES AND SIZES Both Clouds have turned out to be much larger than was once thought. Modern estimates put them at a distance of about 180,000 light-years from the sun, or three times Shapley's earlier estimate. They are also known to extend well beyond their naked-eye limits, for when the outlying Cepheids and clusters are included, their apparent sizes are almost doubled. Radio observations at a wavelength of 21 centimeters show that they exceed even the optical limits, for they are embedded in one great complex of neutral hydrogen. The diameter of the Large Cloud is therefore at least 40,000 light-years, or nearly one half that of the galaxy, while the diameter of the Small Cloud is approximately 25,000 light-years.

STRUCTURE John Herschel and other early observers thought it remarkable that the limited space of the Large Cloud should contain so many stars, star clusters, and nebulae. These objects, they suggested, formed an organized whole, but apart from its dense broad bar, the Cloud showed no signs of structure. It is difficult even today to visualize the Clouds as flattened systems of stars and gas, and still more difficult to determine their orientations in space. What little data we have indicates that they are flattened rotating systems like our galaxy. For some years they were thought to be typical examples of irregular galaxies, or systems with no conspicuous symmetry of rotation and no nuclei. Although this still applies

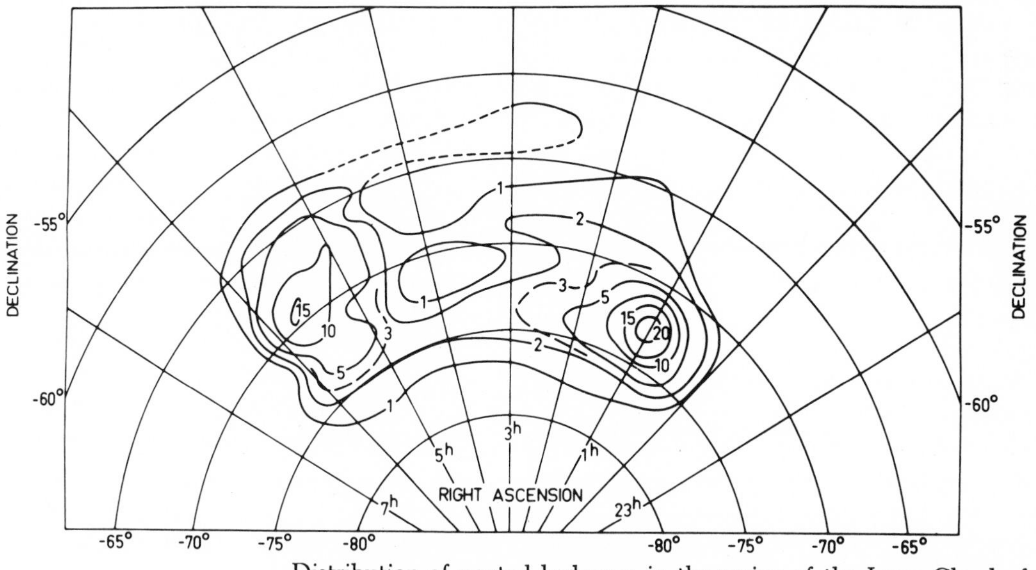

Distribution of neutral hydrogen in the region of the Large Cloud of Magellan (left) and Small Cloud of Magellan (right). The contours of relative intensity show that the Clouds are embedded in a great complex of neutral hydrogen.
—DIVISION OF RADIOPHYSICS, CSIRO, SYDNEY, AUSTRALIA

to the Small Cloud, the Large Cloud may be an irregular barred spiral. Star counts by G. de Vaucouleurs of the University of Texas show that there are good reasons for regarding the Large Cloud as a flattened system with a very small nucleus crossed by a broad bar. An arm curves away from each end of the bar to give the system an S-shaped structure, but the arms are fragmentary and difficult to trace. In addition, the Cloud is foreshortened, its plane being inclined at an angle of 30 to 45 degrees to our line of sight.

The Clouds are well clear of the galactic plane and therefore fairly free from the effects of galactic haze. Even so, they are literally peppered with foreground stars, and the ability to distinguish these from those that actually belong to the Clouds is still a big problem. The mass of the Large Cloud is probably about a tenth that of the galaxy, and that of the Small Cloud a thirtieth, so each system contains at least many millions of stars. Yet a star as luminous as the sun would be of apparent magnitude 23 and therefore difficult to detect even in a 200-inch telescope. Several RR Lyrae stars have been found in globular clusters in both Clouds, but even these are only of magnitude 19 and therefore just within the reach of a 74-inch telescope. It follows that until a really large telescope is available in the Southern Hemisphere, our study of the stellar contents of the Clouds will have to exclude nearly all the main-sequence stars.

SUPERGIANT STARS IN THE LARGE CLOUD The brightest stars in the Large Cloud are blue-white supergiants with apparent magnitudes between 9 and 10. Admittedly they are faint, but when we consider their immense distance, they put most of our galactic supergiants completely in the shade. The faintest are still 100,000 times more luminous than the sun, while S Doradus, of average absolute magnitude about −8 and one of the brightest of the blue supergiants, approaches the million mark. The latter, a member of the large open cluster NGC 1910, is also an intrinsic irregular variable of a type similar to Eta Carinae. Minima occurred in 1891, 1900, 1930, 1940, and 1964, the last being magnitude 10.9, the lowest ever recorded. Stars like this must have enormous masses and comparatively short lifetimes.

GLOBULAR AND OTHER STAR CLUSTERS Many of the highly luminous stars in the Clouds are found in globular clusters and clusters of the galactic type. The former have two remarkable peculiarities. First, they do not form a "halo" around either Cloud, and second, they appear to be in all stages of evolutionary development. In the Large Cloud, for example, about thirty-five globular clusters resemble those associated with the galaxy. Their red giants and supergiants are rich in hydrogen and poor in metals, but even the oldest among them contain stars that are slightly bluer than their "normal" galactic counterparts. A similar number of other clusters look like globulars but their brightest stars are not red but blue. Some of them even contain classical Cepheids of Population I, an association completely unknown in our galaxy. They could, of course, be recently evolved clusters, but there is evidence that they have peculiarities in chemical content due (presumably) to the composition of the material in which they were born.

Clusters of young blue giant and supergiant stars are particularly numerous in the Large Cloud: nearly five hundred have been found in the densely populated central parts alone. Some have the compact appearance and circular outlines of globulars, others are irregular and fairly open. They represent the largest and brightest members of what is probably a very large cluster population, but the majority look quite tiny even on large-scale photographs. We have constantly to remind ourselves of their great distance. At 180,000 light-years a cluster like the Pleiades would be no brighter than an object of the fifteenth magnitude and hardly discernible as a cluster.

Some of the brighter clusters declare their youth by their association with glowing clouds of hydrogen gas. The most oustanding example is the 30 Doradus complex with its central cluster of blue supergiants. The latter is so rich in ultraviolet light that it excites the gas clouds into luminosity out to a distance of about 400 light-years. This magnificent nebula is thus some 32 times more extensive than the Great

Nebulous complex around 30 Doradus in the Large Cloud of Magellan. Also known as the Tarantula Nebula. Photographed in red light.
—Mount Stromlo Observatory

Nebula in Orion and its mass is believed to be equal to that of 500,000 suns. It is estimated that if 30 Doradus were as near to us as the Orion Nebula it would easily cover the entire constellation of Orion. Its total brightness would be two or three times that of Venus at its brightest, and it could easily cause objects at night to cast perceptible shadows.

Shapley's "Constellations" The Large Cloud also contains very large star groups similar to the O and B associations found in the galaxy. Shapley called these assemblages "constellations" and found that their brightest members were mainly blue giants and supergiants. A good example is "Constellation I," or NGC 1935. Here several hundred stars, of total mass 24,000 suns, are enmeshed in a great cloud of ionized and neutral hydrogen some two hundred times more massive than all the stars combined. This is enough gas to maintain the present rate of star formation for many millions of years. The association is obviously a young one, but it is not known for certain whether it is expanding. At its distance a velocity of expansion of 5 miles a second would give rise to a maximum proper motion of only 1 second of arc in 36,000 years.

NEBULAE AND NEBULOSITY Also included in the Large Cloud are dark nebulae, streaks and wisps of bright nebulosity, planetary nebulae, Wolf-Rayet stars, an immense number of classical Cepheids, and numerous variables of the long-period irregular and eclipsing types. Several novae have been discovered in both Clouds: their maximum luminosities were comparable to those of novae in our galaxy. No supernovae have been recorded, but in the Large Cloud three supernovae remnants have recently been discovered as a result of the extremely high intensities of their radio emissions. From these observations we can conclude that the Large Cloud contains an assortment of objects similar to those found in the galaxy.

Much the same can be said of the Small Cloud, although this shows less bright nebulosity and its young blue stars, although plentiful, are less predominant. The Cloud is also comparatively free from interstellar dust, for very remote galaxies, looking like tiny patches on long-exposure photographs, can be seen in the background. These and other differences indicate that the stage of stellar evolution is more advanced in the Small Cloud than in the Large Cloud.

DISTRIBUTION OF HYDROGEN GAS Studies of the Clouds at radio wavelengths took a big step forward in 1961 when a 210-foot steerable radio telescope was completed at the National Radio Astronomy Observatory at Parkes, near Sydney, New South Wales. With this it is possible to plot the intensity distribution of hydrogen in considerable detail over a large area and to achieve a degree of resolution of about 5 miles a second in radial velocity. Two emissions are received—one at the 21-centimeter wavelength due to neutral hydrogen and another, continuous and thermal in origin, due to ionized hydrogen.

Large concentrations of hydrogen are found in the Large Cloud, often in association with groups of hot stars and emission nebulosities. There are also well over two hundred highly localized or discrete areas of ionized hydrogen, many of which coincide with bright emission nebulae. In the Small Cloud, where the hydrogen is distributed more uniformly, there are comparatively few individual concentrations.

RELATIVE AGES OF THE CLOUDS The early belief that each Cloud was an organized whole is now an established fact, but the degree of organization is nowhere near as advanced as that in the galaxy. The irregular structure of the Large Cloud, its high hydrogen content (about 7 percent compared with 2 percent for the galaxy), the large number of giant blue stars, and the presence of globularlike clusters of blue stars are all characteristics of youth. Much the same is true of the Small Cloud, although certain population differences already mentioned give the system a slightly older appearance. But signs of youth do not imply youth. The Clouds could be as old as the galaxy and owe their apparent youth to differences in original composition and to slower rates of evolution.

Although the Clouds move in the gravitational field of the galaxy it is not certain whether they travel about it like satellites. Radial velocity measurements show that the Large Cloud is receding from the center of the galaxy at an average velocity of 23 miles a second and that the Small Cloud is approaching the center at a velocity of 10 miles a second. It is highly probable that they revolve about one another and also about the galaxy for they are certainly not satellites in the sense that the galaxy is a giant and they are dwarfs. We shall see presently that the three systems, along with at least seventeen others, form a "Local Group" or "Local Cluster" in which the Clouds are surpassed in size and mass only by M31, our galaxy, and M33.

M31 in Andromeda

M31, the Great Galaxy in Andromeda, is the only other galaxy visible to the unaided eye. In the sky it looks like a hazy oval-shaped patch about midway between the Great Square of Pegasus and the W formed by the five brightest stars of Cassiopeia, and it is well placed for observation at northern observatories, especially those in mid-latitudes where it passes almost directly overhead. Simon Marius, who in 1612 first saw it through a telescope, thought that it resembled a candle flame seen through horn. Messier represented it as two cones of light joined at their bases. William Herschel wrote: "The brightest part of it approaches to the resolvable nebulosity, and begins to show a faint red color," but his son John considered that its light was "of the most perfectly milky, absolutely irresolvable kind."

In 1864, Huggins, encouraged by his discovery of the gaseous nature of the planetary nebula in Draco, turned his telescope and spectroscope toward M31. He saw, much to his surprise, an extremely faint continuous spectrum, although he realized that this in itself did not prove that M31 was composed of stars. The solution came in 1899 when J. Scheiner at the Potsdam Astrophysical Observatory photographed the spectrum with an exposure time of 7½ hours and obtained definite evidence of dark lines similar to those in the sun's spectrum. He accordingly announced that the nebula was an immensely distant cluster of stars, an interpretation that was widely accepted until 1912 when Slipher found that the reflection nebulosity in the Pleiades gave a similar spectrum.

Meanwhile M31 firmly resisted all attempts to resolve it into stars. In 1888 an English amateur astronomer named I. Roberts obtained photographs which revealed its spiral structure. The discovery came as no great surprise, since Lord Rosse, using a 72-inch reflector some 38 years earlier, had detected the spiral structures of M51 and M99 in Virgo. Astronomers naturally surmised that M31 owed its oval appearance to foreshortening—that in reality it was a round and fairly flat nebula of the "whirlpool" type. Early photographs also showed that it extended well beyond its visual limits and covered an area greater than that of the Large Cloud of Magellan. But since its distance was unknown there was nothing to suggest that it was comparable to the galaxy in size. Miss A. Clerke, an English astronomer writing in 1890, quoted 65 light-years as a reasonable guess of the distance and thought that its diameter of 6 light-years was incredibly large.

DISTANCE An indication of its great distance came in 1885 when a "new" star flared up near its nucleus. The star, known as S Andromedae, attained magnitude 7.2 at maximum and for a few days contributed considerably to the total brightness of the spiral. Unfortunately the distances and therefore the absolute magnitudes of other (galactic) novae were then unknown, but there could be no doubt that S Andromedae owed its faintness to its very considerable distance.

The first photographs to show the brighter stars of M31 as individual objects were taken in the early 1920's with the 100-inch Hooker reflector of the Mount Wilson Observatory. As astronomers examined the outer parts of the image of the spiral they could just make out star clouds, star clusters, and emission nebulae. Then, by comparing photographs taken at different times, they were able to detect several novae, and most important of all, a Cepheid-type variable. The more they studied the photographs the more they got the impression that they were looking at their own galaxy from afar. The spiral had every indication of being an immense stellar system, or "island universe," far beyond the confines of the galaxy. And its distance? The solitary Cepheid, detected toward the end of 1923 by Hubble, had a period of about a month, and according to the period-luminosity relationship, an absolute magnitude of −4. Its apparent magnitude was only 18.2 at maximum, so to appear as faint as this its distance had to be at least 900,000 light-years.

This result, needless to say, required urgent confirmation, so M31 was immediately submitted to a most detailed and extensive study. As this proceeded the Cepheids grew in number, various irregular variables were discovered, more star clusters and patches of diffuse nebulosity were identified, blue giant and supergiant stars showed up in great profusion along the spiral arms, and novae appeared at a rate of about eight a year.

The Cepheids supported the provisional finding of 900,000 light-years, so also did the novae, for it was assumed that their average maximum absolute magnitude was about −6, or roughly that of novae in our galaxy. S Andromedae was a conspicuous exception. Its brightness had been comparable with that of the whole nebula, but the other novae were all faint objects and ranged in apparent magnitude from 16 to 18. Baade of the Mount Wilson Observatory then concluded that there were two groups of novae differing in luminosity by a factor of the order of 10,000. In other words, S Andromedae, with an absolute magnitude of −15 at its brightest, was a supernova—a stellar explosion so stupendous that it could be seen with a pair of binoculars over a distance then reckoned to be some 5 million million million miles.

LINEAR DIMENSIONS At this distance, and with 3 degrees for its apparent diameter, M31 acquired a linear diameter of 50,000 light-years. This meant that it was a mere island compared with the continent of Shapley's galaxy. After 1930, however, and for reasons given in the previous chapter, the diameter of the galaxy shrank to about 100,000 light-years. The disparity in size was reduced still further when photoelectric studies showed that M31 had an overall length of at least 5 degrees. Its linear extent accordingly almost doubled and it became comparable in size to our galaxy.

In 1942 the photography of M31 and other galaxies was still being done with blue-sensitive plates. Baade, working with the 100-inch telescope during the war years, found that the blackout imposed on Los Angeles about eighteen miles away enabled him to record stars fainter than ever before. Some of his photographs of M31 showed signs of stars as faint as photographic magnitude 21, but the nucleus still looked diffuse and structureless. He then decided to try red-sensitive plates used in conjunction with filters which limited the light to a fairly narrow band near the red end of the spectrum. This completely changed the situation, for the nucleus was resolved into a structureless mass of myriads of stars.

BAADE'S CONCEPT OF TWO STELLAR POPULATIONS These results led Baade to introduce the concept of two types of stellar population. He realized that the stars in M32, NGC 205, and the nucleus of M31 were similar to the brightest stars in globular clusters but different from those in galactic clusters and in the solar neighborhood. At the time, the distinction rested on the relative numbers and space distribution of different types of stars. Other characteristics, such as differences in the chemical composition of stellar atmospheres, came later.

For many years some astronomers had a strong suspicion that something was wrong with the calibration of the period-luminosity relationship. It had been used to determine the distances of several nearby galaxies, yet the novae that appeared in them had calculated luminosities invariably greater than those of novae in our galaxy. Similar differences, of one or two magnitudes, were found with globular and open clusters. The situation was not helped when the estimate of the distance of M31 was reduced to allow for the effects of galactic absorption. At the new value, 750,000 light-years, any RR Lyrae stars in M31 should have been within reach of the 200-inch telescope on Palomar Mountain, but a careful search by Baade in 1954 failed to reveal any. On the other hand, if one based the distance of M31 on its novae and globular clusters, its RR Lyrae stars would be too faint to be detected with the 200-inch telescope.

TWO TYPES OF CEPHEIDS Other studies confirmed and extended these findings and showed where the fault lay. There were two types of Cepheids. The classical Cepheids, Type I, were members of Population I and typical of the arms of spiral galaxies. The Cepheids observed in globular clusters, Type II, were members of Population II and typical of the central regions of spiral galaxies. Further, for any one particular period of light variation, a Type I Cepheid was more luminous than one of Type II by about 1.5 magnitudes, that is, by a factor of 4. As a result the distances of the Clouds, M31, and other nearby galaxies had to be doubled, for they were based on Type I Cepheids. M31 straightway shot out to a distance of 1.5 million light-years (about three fourths of the present value of 2.2 million light-years) and its overall diameter became 180,000 light-years, or almost double that of our galaxy.

SPIRAL ARMS OF M31 M31, like our galaxy, has two spiral arms. They curve away from the nucleus, and on photographs taken in the red light of hydrogen, can be traced for at least two coils. They probably make three or more complete turns before they finally trail off into space, but their outermost parts are beyond the photographic limits of the 200-inch telescope. Their general plane, that of the galaxy itself, is tilted by about 15 degrees to the line of sight, so for the most part they appear highly foreshortened. This would make them very difficult to trace on ordinary photographs were it not for outlining lanes of interstellar dust. The latter, reminiscent of the dark nebulae near Rho and Eta Ophiuchi, are most noticeable in the inner regions. Farther out, where luminous blue stars, Cepheids, and emission nebulae are abundant, the dust clouds become less noticeable and finally disappear altogether. In the outermost regions the general substratum of dust is so thin as to enable more distant galaxies to be seen through gaps in the spiral arms.

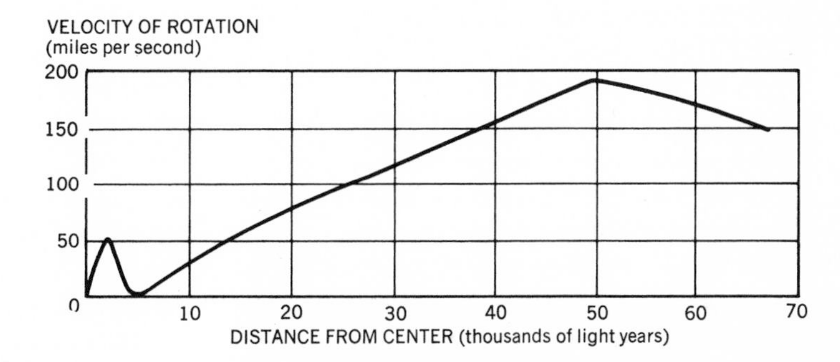

The rotation of M31 according to observations by M. Humason and N. U. Mayall. The system rotates like a solid body from the center to a distance of about 2,500 light-years from the center, and also between the limits 5,000 and 50,000 light-years. Beyond 50,000 light-years the rotation is Keplerian, similar to planetary motions in the solar system.

ROTATION By measuring the Doppler displacements of the narrow lines of emission nebulae in the spiral arms, Humason and Mayall have found that the angular rotation of M31 is fairly constant over the range 5,000 to 50,000 light-years from the center. Between these limits the system rotates like a solid body, so its density must be roughly uniform. In this respect it is like our galaxy, although the maximum orbital motion (190 miles a second at 50,000 light-years from the center) occurs at twice the corresponding distance for our galaxy. This is not surprising since the overall diameter of M31 is nearly twice that of our galaxy. Beyond 50,000 light-years the orbital motion, and therefore the density, gradually decreases with increasing distance. Similar results have been obtained for M33 and several other spiral galaxies: in general their inner parts (the nuclei excepted) have fairly uniform rotations (and therefore mass distributions) out to about 0.4 of their radii.

MASS If the velocity of rotation of a galaxy at a specific distance from its center is known, it is possible to estimate its total mass. That of M31 is of the order of 200 billion suns, or roughly twice that of our galaxy. Most of this is in the form of stars, for 21-centimeter radio observations indicate that hydrogen gas accounts for only 1 or 2 percent of the total. Yet this small percentage could form at least a billion stars, each of them equal in mass to the sun.

TWO COMPANIONS—M32 AND NGC 205 M31 resembles our galaxy in other ways. It has its own system of about two hundred globular clusters, is surrounded by an enormous spherical "corona" of radio emission, contains planetary nebulae, has concentrations of neutral hydrogen in its spiral arms, and owes most of its light to its highly luminous Population II stars. It also has two companion galaxies, M32, first described by Le Gentil in 1749, and NGC 205, discovered by William Herschel's sister, Caroline, in 1783. These companions differ completely from the Clouds of Magellan. They have regular elliptical outlines, but no spiral or other internal structure, and practically no interstellar gas or dust. Baade, using the 200-inch telescope and red-sensitive plates, found that their brighter stars were red giants and supergiants of Population II. They resemble, although on a greatly reduced scale, the struc-

M33 in Triangulum, a nearby spiral galaxy, photographed in red light and showing partial resolution into stars.
—Mount Wilson and Palomar Observatories

A faint, dwarf galaxy in Sextans.
—Mount Wilson and Palomar Observatories

tureless Population II systems of the galaxy and M31 itself.

When the 200-inch telescope is used to study the system of M31 its powers of revealing detail are equivalent to those of a 20-inch telescope trained on one of the Clouds of Magellan. This unfortunate situation is due solely to the great difference in distance between these galaxies. It follows that dwarf stars in M31 are at present quite beyond detection: the apparent brightness of a star like the sun would be of the order of magnitude 29. By the same token our knowledge of all but a few other galaxies is even more restricted: M31 and its companions, despite their great distance, are practically on our doorstep.

M33 in Triangulum

Another nearby galaxy is M33 in the little constellation of Triangulum. A spiral system, its plane is almost at right angles to the line of sight, so the nuclear regions and arms are seen to advantage. Its distance, based mainly on Cepheids and novae, is approximately 2 million light-years and its overall diameter is about 50 thousand light-years. It is therefore as remote as M31 but much smaller in size.

M33 has loosely coiled spiral arms. They curve away sharply from the nucleus in two broad streams of stars, gas, and dust, and can be traced to about one half of a complete turn. Numerous patches of emission nebulosity, together with groups of highly luminous blue stars and extensive star clouds give them a decidedly fragmented appearance. These Population I features, unlike those of M31, can be traced right up to the nucleus. The latter, in fact, is comparatively small and on most photographs looks like a bright knot or condensation in the middle of what is basically an S-shaped system. Photographs in red light with the 200-inch telescope have failed to resolve the main mass of the nucleus into stars, so Population II objects may be relatively scarce.

The central part of M33 rotates almost like a solid body, but beyond 4,000 light-years from the center the motion is similar to that of the planets about the sun. The maximum velocity of rotation (at about 4,000 light-years) is 75 miles a second, and the total mass is believed to be of the order 5 to 8 billion suns. Although the latter is much smaller than that of our galaxy, M33 is still a comparatively large and massive system. In the Local Group it is surpassed only by M31 and our galaxy. Four of its neighbors (the Clouds, NGC 6822, and IC 1613) are irregular systems, and ten are small, compact, elliptical galaxies similar in outline to M32.

Dwarf Galaxies

In 1938 Shapley discovered two small elliptical galaxies similar to the companions of M31. He called them the Sculptor and Fornax superclusters, since they were found in the constellations of those names and looked like enormous but sparsely populated globular clusters. Further study showed that they were members of the Local Group and that their extreme faintness arose from their low luminosities. The Fornax system, about 600,000 light-years away, has a diameter of the order of 22,000 light-years, while the one in Sculptor, about 270,000 light-years away, is roughly 7,000 light-years across. Their masses, however, are similar to those of the larger globular clusters associated with our galaxy, so their stars form exceptionally loose aggregates.

More nearby dwarf galaxies were discovered after 1948 as a result of extensive photographic surveys made with the 48-inch Schmidt telescope on Palomar Mountain. Of these the smallest is a faint elliptical system in Ursa Minor with a diameter of about 3,000 light-years. Its stars are so widely spaced that their density even at the center is reckoned to be less than a thousandth of the star density in the neighborhood of the sun. But not all dwarf galaxies are ellipticals. NGC 6822 and IC 1613 are quite ragged and irregular, and unlike the dwarf ellipticals, contain large numbers of hot, blue, and therefore young, stars.

THE EXTENT OF THE LOCAL GROUP The Local Group occupies a region of space about 3 million light-years in diameter, with our galaxy and M31 on opposite sides of the center. The shape of this region is ellipsoidal rather than spherical: the minor axis is about 1½ million light-years and the depth or thickness is roughly a quarter of this value. Since galaxies are large objects compared with the distances between them, this volume of space is more densely packed than even the most crowded star cluster. A few coins of various sizes, suspended over the top of a breakfast table, would, at a pinch, represent the distribution of the galaxies in the Local Group. But if we represented a single star by a dime, a neighboring star would be another dime several hundred miles away.

CHAPTER XIII

THE UNIVERSE OF GALAXIES

The members of the Local Group are the only galaxies whose stellar contents can be studied in any great detail. In more distant systems, the Cepheids can be too faint for detection even with the 200-inch telescope, and distances have to be based on novae and the very brightest supergiants. For galaxies beyond 20 million light-years, the estimated distances usually depend on observations of supernovae and the apparent brightnesses of the systems themselves. Supernovae are unfortunately all too rare and apparent brightness depends on both luminosity and distance.

Galaxies, like stars, have a considerable range in luminosity. The absolute magnitude of M31, for example, is −21.2, but that of the dwarf elliptical in Ursa Minor is about −9. M31 is, therefore, more than 120,000 times more luminous than its small neighbor.

Hubble's Pioneer Investigations

The fact that the distance of a galaxy cannot be based solely on its apparent brightness was known to Hubble in the 1920's. He attempted to overcome the difficulty by determining the average luminosities of groups of galaxies. He argued that, although individual galaxies had a considerable range in luminosity, the averages of large groups, chosen at random, ought to be fairly constant. This constant, known as the "luminosity function," was then applied to galaxies in 1,283 separate regions, scattered in a fairly uniform manner over 75 percent of the sky.

Hubble's surveys showed that galaxies had an approximately uniform distribution (except, of course, in the region of the Milky Way, where the general galactic haze and individual dust clouds markedly reduced their numbers). More important, he found that the large-scale distribution in depth in space was also fairly uniform. The number of galaxies increased, not only with the limits of increasing faintness but also in direct proportion to the volume of space surveyed.

"Thus the observable region," he wrote in 1936, "is not only isotropic but homogeneous as well—it is much the same everywhere and in all directions. The nebulae [galaxies] are scattered at average intervals of the order of two million light-years or perhaps two hundred times the mean diameters. The pattern might be represented by tennis balls fifty feet apart."

Hubble estimated that galaxies of magnitude 21, the faintest ever recorded with the 100-inch telescope, were at an average distance of about 500 million light-years. They marked the frontiers of a universe that contained about 100 million galaxies "in various stages of their evolutionary history," and which had, by definition, our galaxy at its center.

Groups and Clusters of Galaxies

Hubble investigated the small-scale distribution of galaxies and found that they had a tendency to form groups and clusters. Systems of two and three galaxies in close proximity were quite numerous. There were also larger groups, analogous to the sparser, open star clusters, and about twenty really large groups or clusters, each consisting of several hundred members. Hubble thought that clusters of galaxies were the exception rather than the rule—that the great majority of galaxies were "field" rather than "cluster" objects. Subsequent surveys, in particular those carried out at Lick and Palomar, have completely changed the picture. It is now pretty clear that clusters of galaxies are the rule and that field galaxies are the exception. There are also traces of an even greater organization, for some groups and clusters appear to form "superclusters," or clusters of clusters of galaxies.

THE CLUSTER IN VIRGO This is the first of the great clusters and it is estimated to be some 30 million light-years distant. Known as the Virgo cluster, its galaxies form a loose aggregate that shows no central concentration or spherical symmetry. Its largest and brightest members lie partly in Virgo and partly in neighboring Coma Berenices and include several Messier objects. Just how far the cluster extends is not definitely known, but it may have an overall depth of about 15 million light-years and an overall length of at least 30 degrees, corresponding to 20 million light-years. Perhaps two clusters are involved, one behind the other, and the system may be associated with other groups in Virgo and Coma Berenices to form part of what Vaucouleurs has termed a "Local Supercluster" or "Local Supergalaxy."

Some of the Virgo galaxies occur in pairs. Their individual masses can therefore be determined in much the same way as can stellar masses, *i.e.*, from the motions of the components of double stars. Several of the spirals can be partially resolved with the 200-inch and other large telescopes. Thus they present opportunities for a comparison of their stellar contents with those of members of the Local Group. Among them are M90, a handsome spiral with distinct whorls of emission nebulae and masses of stars; M60, a tightly wound spiral; and M100, a spiral well-tilted to the line of sight, but showing numerous patches and lanes of dark nebulosity.

Also in Virgo, although not necessarily a member of the cluster, is M104 (the Sombrero Galaxy), a spiral that is seen edgeways. Its stars cannot be resolved, but photographs in red light clearly show the almost spherical distribution of Population II, while those in blue and ultraviolet bring out the more disklike distribution of Population I. A striking feature is a very dark, intense band that runs right across the nucleus and along the periphery of the spiral arms, thereby identifying the nearer edge of the system.

Among the elliptical galaxies in the Virgo cluster is M87, an enormous globular system of Population II stars with a total mass thirty times that of our galaxy. On photographs, it has a completely amorphous appearance and is surrounded by scores of hazy specks, undoubtedly globular clusters too distant to be resolved into stars.

At distances greater than 100 to 200 light-years, small groups like the Local Group would be difficult to recognize, and we are left with large and rich aggregates which show up by virtue of their immense size. G. O. Abell, from a study of photographs taken with the 48-inch Schmidt telescope at Palomar, has identified no less than 2,712 clusters, each containing more than fifty members. Some, like the Virgo cluster, are irregular, while others show marked central concentrations and spherical symmetry. Among the irregulars is the Hercules cluster at an estimated distance of 330 million light-years. Within its limits, over 1,000 galaxies of practically all types have been observed—with the exception, of course, of dwarfs, which are too faint to be seen. Examples of the regulars are the extremely rich clusters in Coma Berenices (300 million light-years away) and Corona Borealis (1 billion light-years away). They contain literally thousands of galaxies, but most, if not all, are ellipticals.

The discovery of so many clusters does not in the least upset Hubble's conclusion that the large-scale distribution of matter is fairly uniform. As ever larger volumes of space have been surveyed, more clusters have been observed; they appear to be as plentiful in one direction as in another. The frontiers of the visible universe have been pushed well beyond Hubble's limit of 500 million light-years.

At present the calculations of the distances to remote galaxies are not precise. For this and other reasons that will be given in the next chapter, astronomers are reluctant to assign distances to those galaxies that are at present detectable with the 200-inch telescope, but a value of the order of 5 billion light-years would probably be fairly accurate. In a universe of this radius, a universe centered on our galaxy by definition, there are believed to be several billion galaxies.

Central region of the cluster of galaxies in Hercules.
—MOUNT WILSON AND PALOMAR OBSERVATORIES

Classifying Galaxies

Another important result of Hubble's surveys was the discovery that the great majority of galaxies form a definite sequence when classified according to their form. Hubble represented the sequence by a diagram shaped like a tuning fork. Elliptical galaxies form the stem or "early" end. Designated by the letter E, they progress from those with globular forms (E0), through ellipticals of increasing flatness (E1 to E7) to disk-shaped systems (S0) that lack spiral arms.

The sequence then divides into two branches. One contains normal spirals (S) whose arms are tightly wound (Sa), less-tightly wound (Sb), and quite open (Sc); and the other contains barred spirals (SB) arranged in a similar sequence (SBa, SBb, SBc). In both cases the transition from a to c is one of increased arm development and openness of structure at the expense of the prominence of the nuclear regions.

Although most galaxies can be assigned to a place in the sequence, a few appear to be misfits. Among the latter are spirals of mixed type; their nuclei have barred characteristics, but otherwise they resemble normal spirals. Hubble placed them between the two branches and obtained some consolation from the fact that they appeared to be comparatively rare. Other misfits are the irregulars, which obviously have no place at all in a sequence of regular forms. The Clouds of Magellan, however, along with a few others, have stellar contents similar to those of open spirals and can therefore be regarded as representing the final or "latest" stage in the sequence. But the others are so

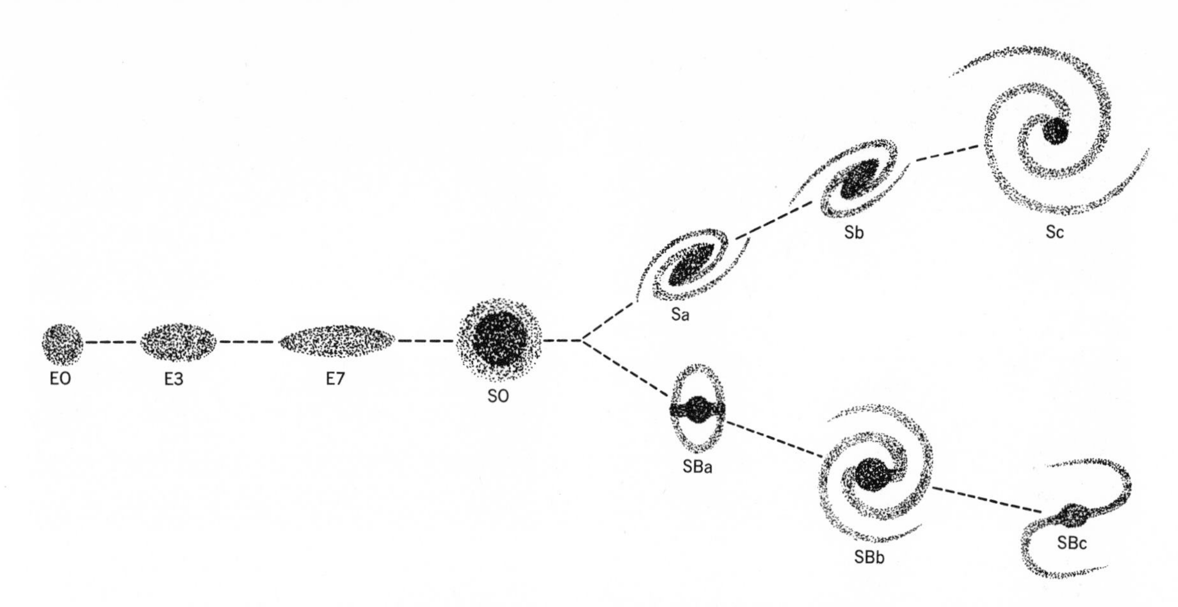

Classification of galaxies according to Hubble. E denotes ellipticals, S normal spirals, and B barred spirals.

peculiar that Hubble had no alternative but to omit them.

M87 in Virgo is a good example of type E0, M32 in Andromeda represents type E2, and NGC 3115 in Sextans represents type E7. M104 in Virgo and M81 in Ursa Major belong to type Sa; our galaxy and M31 are typical Sb; and M33, M51, and M74 are Sc. Examples of barred spirals are NGC 2859 (SBa), NGC 1300 (SBb), and NGC 1073 (SBc). Although Hubble regarded type S0 as a "more or less hypothetical class," galaxies of this kind do exist, *e.g.*, NGC 5195, the small companion of M51, and NGC 7332 in Pegasus.

Spectral Types Another, although less definite, way of classifying galaxies is to arrange them in groups or types according to their spectra. The spectrum of a galaxy is a composite affair that depends mainly on the spectral types in its population. It also depends on the particular wavelength region selected; for the purposes of comparison, astronomers always work within well-defined wavelength limits. They then find that there is a definite correlation between form and spectrum, and, therefore, with stellar population. In general, elliptical galaxies are of type G4, and the spirals range from G3a, through G2b, to F0c.

Evolutionary Trends? Hubble labeled a, b, and c spirals "early," "intermediate," and "late," respectively. He realized that form had evolutionary significance, but resisted any attempt to give the sequence an evolutionary interpretation. When he called a galaxy early or late, he was merely referring to its position in a morphological classification. Some less cautious astronomers thought that the diagram was nothing less than an evolutionary tree. A galaxy, they suggested, started off as a spheroidal mass of stars. It then gradually flattened as its rotation increased and grew into a disk with arms that became more open and prominent.

The spectral trend, however, is precisely opposite to what one would expect if the course of evolutionary development were from ellipticals to spirals. A young galaxy, with plenty of hot, blue supergiants and emission nebulae, would tend to have an early-type spectrum (O, B, A, or F), whereas an old one, with a preponderance of red giants and supergiants, would tend to have a redder or later-type spectrum. It follows that a more probable evolutionary sequence would be one from irregulars to spirals of types c, b, and a. If this is so, the forces operating in a galaxy tend to produce symmetry from dissymmetry, or order from disorder.

Until recently it was thought that if a spiral evolved in an undisturbed state, it would automatically acquire the characteristics of an elliptical. On this basis, spirals and ellipticals could be expected to have a similar range in size and

NGC 1201 Type S0

NGC 2841 Type Sb

NGC 2811 Type Sa

NGC 3031 M81 Type Sb

NGC 488 Type Sab

NGC 628 M74 Type Sc

Different types of normal spiral galaxies.
—Mount Wilson and Palomar Observatories

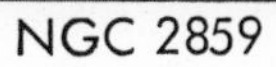

NGC 2859 Type SB0

NGC 2523 Type SBb(r)

NGC 175 Type SBab(s)

NGC 1073 Type SBc(sr)

NGC 1300 Type SBb(s)

NGC 2525 Type SBc(s)

Different types of barred spiral galaxies.
—MOUNT WILSON AND PALOMAR OBSERVATORIES

mass. But, while spirals have a small range, ellipticals have a large one. How, then, can we account for the existence of small ellipticals like M87, one of the companions of M31? Before even a comparatively small spiral could evolve into a so much smaller system, it would have to shed the greater part of its mass. Likewise, a large spiral like M31 would have to undergo something like a thirty-fold increase in mass as it evolved toward the present state of a giant elliptical like M87.

YOUNG AND OLD GALAXIES Some astronomers are of the opinion that galaxies start as gas clouds with different innate characteristics. The mass and angular momentum of a cloud are thought to be largely responsible for the shape of the galaxy formed from it, while the average density determines the rate of star formation. They therefore suggest that galaxies evolve at different rates and in different ways, and cannot be expected to keep to a common evolutionary track.

It has to be realized that a young-looking galaxy, with plenty of hot, blue stars and emission nebulae, may not be young in years. Such a galaxy, whether irregular or spiral, could contain enough hydrogen to support the formation of bright stars over many billions of years. Observations in the 21-centimeter radiation of neutral hydrogen tend to support this idea. Although the results obtained so far are highly tentative, they indicate that the hydrogen content ranges from about 16 percent for irregulars to about 8 percent for well-developed spirals. These are average values; individual spirals appear to contain anything from about 1 to 14 percent.

Of course, the Population II contents of ellipticals brand them as systems well advanced in stellar development, and in this respect they are older than spirals and irregulars. Again, the Population II central regions of spirals indicate that evolution has proceeded further there than in irregulars. But whether all galaxies are of the same age in the sense that they started to evolve at about the same time is an open question. The only galaxies that might have started to evolve at about the same time would be the components of certain physical pairs and small groups but, more often than not, the components are considerably different in type.

INTERACTING GALAXIES Recently, certain pairs and groups of galaxies have received a great deal of attention. Their components either are linked by bridges of luminous material or are so close together that they distort one another. An easily observed example of a connected and presumably interacting pair is M51 and its S0 companion, NGC 5195. Luminous filaments extend from the latter to link with one of the spiral arms of its much larger and brighter associate.

In most other interacting pairs, the visual appearance is more complex. Connecting bridges, like the galaxies they join, show a remarkable diversity of form and structure. Some are long and narrow, others are short and broad. Some are fairly regular in shape and uniform in

The optical part of the intense radio source Cygnus A. One interpretation was that two galaxies were colliding, but it now seems more likely that the object is an exploding galaxy. The redshift indicates a velocity of recession of about 10,600 miles a second.
—MOUNT WILSON AND PALOMAR OBSERVATORIES

brightness, others are highly irregular and divided into knots. Whether they consist of stars, luminous gases, or a mixture of both is still subject to speculation, as is the nature of their origin and the forces that control them. But the very fact that they exist shows that galaxies, like stars, can form close physical associations that must undoubtedly affect their own individual development. This promotes the thought that intergalactic space, instead of being empty, contains material that may have an appreciable absorbing effect on the light of extremely remote galaxies.

PECULIAR GALAXIES Lately, attention has also been focused on the so-called peculiar galaxies, since many of them have been identified with discrete sources of radio emission. The reader will recall that our galaxy and M31 are radio galaxies in the sense that they emit radio energy by the synchrotron process. M33, and probably most if not all normal spirals, are also sources of radio emissions. The intensity of their emissions is in the order of 10^{28} kilowatts; this is a modest radio power compared with that found among certain other galaxies forming at least three different groups, nearly all of them having some kind of optical peculiarity.

Radio Galaxies

CYGNUS A Prominent in one group is Cygnus A. The great strength of its emission led to its early discovery (in 1947); for a short time it was the only known discrete source in the radio sky. In 1951, Baade and R. Minkowski found that it coincided with the position of a galaxy now estimated to be some 500 million light-years away. Further studies then showed that the galaxy emitted a fantastic 10^{34} kilowatts of radio energy of the synchrotron type, or 1 million times that of a normal radio galaxy. The galaxy itself looked double, as if two nuclei were in contact, and for a time it was thought that two galaxies were colliding. The modern interpretation is that we are looking at a single galaxy whose nucleus is splitting apart, or which has a band of dust across its nucleus. Curiously enough, the radio energy comes not from the galaxy itself, but from two regions some 300,000 light-years apart, on each side of the visible nucleus or nuclei.

CENTAURUS A Another member of this group is Centaurus A, identified with the galaxy NGC 5128, at an estimated distance of about 12 million light-years. The latter looks like a giant spheroidal system crossed by a dark, somewhat ragged absorbing band, presumably of dust. In this case, the radio intensity is about 10^{31} kilowatts, and the emission comes from four regions. Two, which are small and intense, are confined to the disk; the other two are symmetrically placed on each side of the disk. All four lie approximately on a line running through the center of the disk and perpendicular to the dark, absorbing band. A similar system is M84, a giant spheroidal galaxy in the Virgo cluster. It broadcasts at about one thirtieth the intensity of NGC 5128, and its dark central band is visible only on short-exposure photographs.

VIRGO A Typical of another group is Virgo A, now identified with M87. Its radio intensity is about the same as that of NGC 5128, but the emission comes from two concentric sources centered on the galaxy. The inner one is small and intense, the outer is comparatively weak and slightly smaller than the visible disk. Ordinary photographs suggest that M87 is a normal giant spheroidal galaxy, but those taken in short time-exposures reveal a long, narrow jet extending from the nuclear region. The jet probably consists of a stream of high-energy gas and is presumably responsible for the radio emission. A similar jet protrudes far beyond the outer edge of a radio source designated 3C 273 (from its number in the *Third Cambridge Catalogue of Radio Sources*), and twin jets extend from opposite sides of a spiral arm in the intense radio galaxy NGC 4651. In these, and similar cases, the jet or jets have probably resulted from extremely violent explosions in the nuclei.

A third group contains spiral galaxies, normal in appearance but having enhanced radio emissions. The spiral NGC 1068, for example, looks normal but emits 10^{30} kilowatts of radio energy, all of which seems to be concentrated in the nucleus. The latter gives a spectrum crossed by intense, broad emission lines, but this is not necessarily significant. Several other spirals exhibit similar optical spectra, but their radio emission is not enhanced.

M82—AN EXPLODING GALAXY? By far the best example of a large-scale nuclear change is found in M82, a galaxy in Coma Berenices whose position coincides with that of the weak radio source 3C 231. On ordinary photographs M82 looks like a highly foreshortened, irregular galaxy, rich in dust, and with faint nebulous wisps in its outer parts. Yet in 1961, when Sandage photographed it in red light with the 200-inch telescope, it took on an altogether different appearance. Its central region was clearly in the throes of a vast upheaval. The wisps turned out to be part of an immense structure of filaments of ionized hydrogen that extended some 14,000 light-years above and below the plane of the galaxy.

NGC 5128, a peculiar galaxy and intense radio source in Centaurus.
—MOUNT WILSON AND PALOMAR OBSERVATORIES

M87, a giant elliptical galaxy in Virgo. Short-exposure photograph shows nuclear jet.
—Lick Observatory

Spectral measurements then showed that the hydrogen complex was expanding at a rate of about 600 miles per second. It could therefore be reasonably surmised that the filaments were the result of a gigantic explosion which occurred some 1.5 million years ago. But since M82 has an estimated distance of 10 million light-years, the time of the explosion was about 11.5 million years ago.

Other optical studies show that the light from the faint outer filaments has a high degree of polarization. Sandage has therefore suggested that M82 possesses a large-scale magnetic field aligned predominantly along its axis of rotation. He reasoned that the filaments, like those of the Crab Nebula, shine by the synchrotron process, and that the explosion produced large numbers of high-speed electrons, whose energy generates not only the observed radio flux but also enough ultraviolet radiation to ionize the

M82, an exploding galaxy in Coma Berenices, in the red light of hydrogen-alpha with the photovisual image photographically subtracted. The photograph is printed as a negative in order to accentuate the details of the hydrogen filaments.
—Mount Wilson and Palomar Observatories

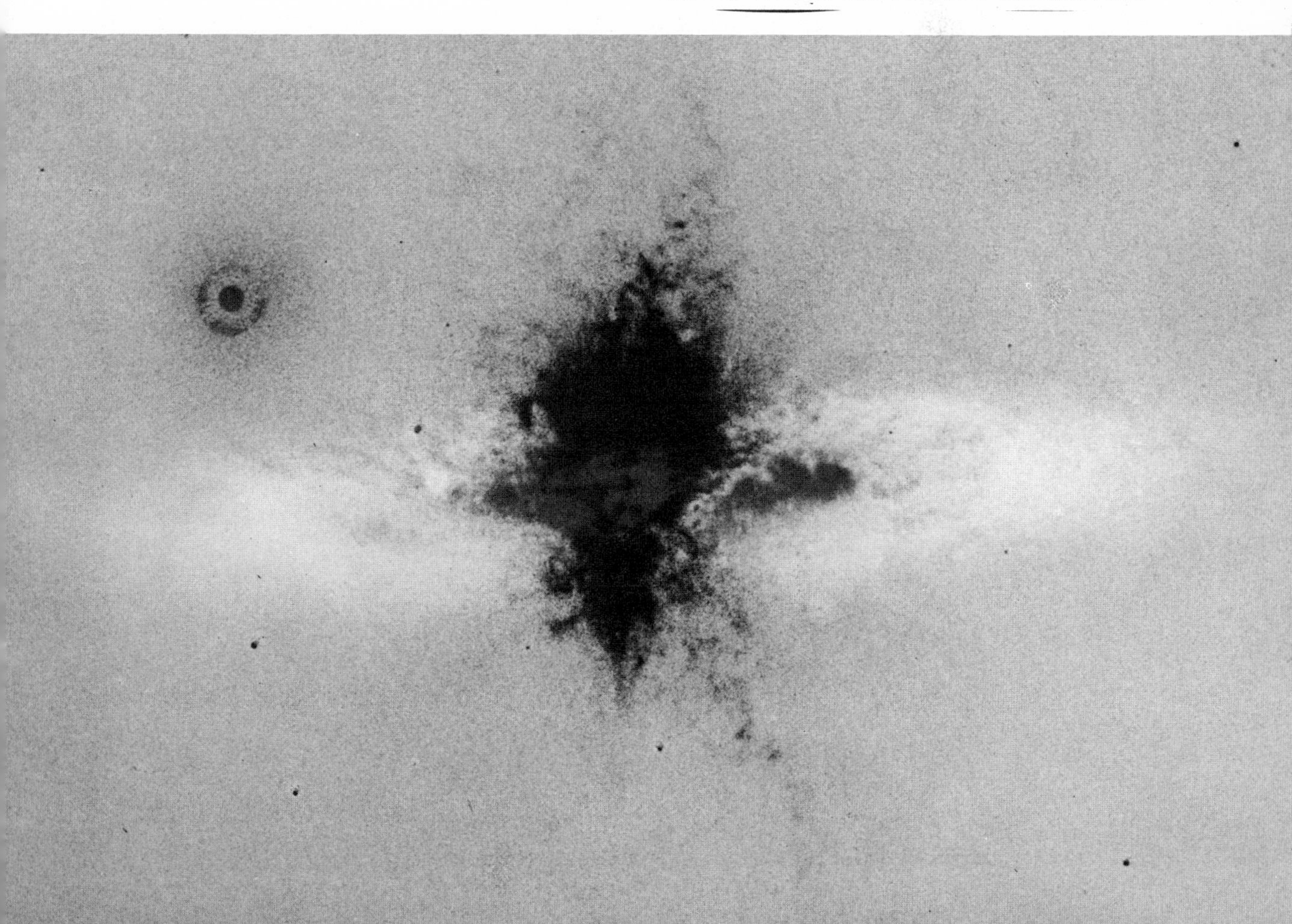

hydrogen in the filaments. According to his calculations, the amount of hydrogen moving away from the center of the galaxy is roughly equal to 5 million times the mass of the sun, or about a two-thousandth of the total mass. An explosion that could set so large a mass in rapid motion would obviously involve energies immensely greater than those released by even the grandest supernova outburst.

Sandage's explanation resolves the puzzle of why the radio emissions of sources like Cygnus A and Centaurus A are concentrated in two spots on each side of the optical object. Presumably two jets of high energy electrons are moving in opposite directions from the center of the parent galaxy and carrying part of the magnetic field with them. As a result, the lines of magnetic force are highly compressed at the ends of the jets, and the synchrotron is subject to a corresponding increase in intensity.

CHAPTER XIV

THE EXPANDING UNIVERSE

RADIAL VELOCITIES OF GALAXIES In 1912, when direct evidence for the extragalactic status of nebulae like M31 and M33 was entirely lacking, Slipher of the Lowell Observatory, Flagstaff, Arizona, made an observation of far-reaching importance. He found that the absorption lines in the spectrum of M31 were all shifted slightly toward the violet end of the spectrum by amounts proportional to their wavelengths. Even more surprising, the shifts, when interpreted as a direct result of motion toward the earth, gave M31 a velocity of approach of about 190 miles a second. This implied that the nebula had a motion all of its own and was therefore extragalactic, for at the time it seemed most unlikely that so high a velocity could be ascribed wholly to the solar system.

In 1925, despite great difficulties in observation, Slipher had obtained the radial velocities of forty-one extragalactic nebulae. Their large range, from −190 miles a second to +1,125 miles a second, established beyond all doubt that these objects were well outside the gravitational control of the galaxy. Allowance had to be made, of course, for the velocity component due to the rotation of the galaxy, but when this was done, the velocities were still exceptionally high. The proper motions of M31 and other galaxies in the Local Group tended to mask their individual recessional motions, but the latter were so large for more distant systems as to show that they were all moving away from the center of our galaxy.

VELOCITY-DISTANCE RELATIONSHIP In 1929 Hubble and Humason began to extend Slipher's work by using the great light-gathering power of the 100-inch telescope on Mount Wilson. More reliable criteria of determining the distances of galaxies had been developed and the motion of the solar system was better known. Hubble soon found that over the observed range of 6.5 million light-years the velocities of recession were approximately proportional to the distances. According to his data, the increase in velocity with distance, known later as the "Hubble constant," was roughly 100 miles a second per million light-years. Here was a quite unexpected way of determining the distance of any galaxy within the observed range.

THE RED SHIFT AND THE HUBBLE CONSTANT The displacement of the spectral lines of a galaxy toward the red end of the spectrum is usually referred to as the "red-shift." The lines in any one spectrum, however, are not displaced by equal amounts. The amount ($\Delta\lambda$) depends on the wavelength (λ) in such a way that the ratio ($\Delta\lambda/\lambda$) is constant. If we assume that the displacements are Doppler effects, the relative velocity of recession (V) is obtained from the *approximate* formula $V = c\Delta\lambda/\lambda$ where c is the velocity of light. According to the velocity-distance relationship $V = HD$, where H represents the Hubble constant and D the distance.

EXTENDING THE VELOCITY-DISTANCE RELATIONSHIP When the data had been checked and rechecked the next step was to test the velocity-distance relation over a much greater distance. This Humason did, using for the purpose special high-speed spectrographs, and with such success that, by 1935, he had added nearly 150 new velocities. Working on isolated galaxies and clusters of galaxies, and using apparent brightness as a criterion of distance, he suspected that Hubble's "law" held good to the extreme limits of observation with the 100-inch telescope. The faintest cluster for which a radial velocity could be obtained had an observed velocity of recession of 26,000 miles a second, or nearly one seventh the velocity of light. On the basis of the Hubble constant the cluster (Ursa Major No. 2) had a distance of about 260 million light-years, a figure which agreed quite well with that based on the apparent brightnesses of

Cluster of galaxies in Corona Borealis with a velocity of recession of 13,500 miles a second.

—Mount Wilson and Palomar Observatories

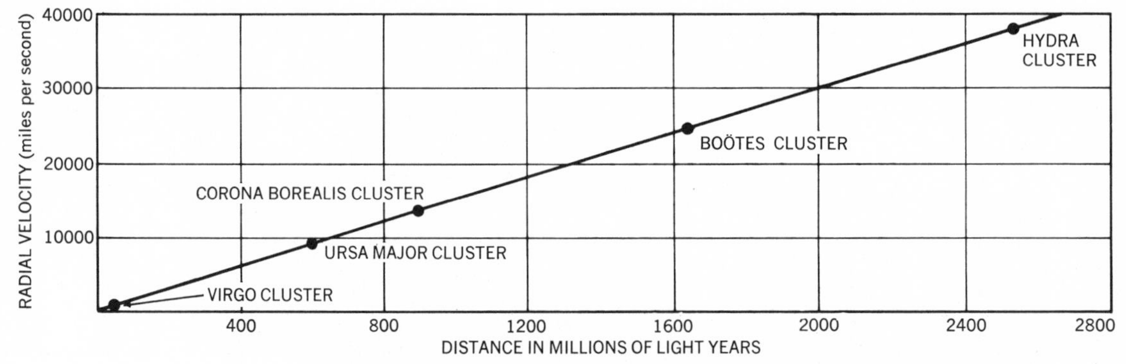

The velocity-distance relationship. The velocities of recession of relatively near galaxies increase in direct proportion to their distance from us. The slope, or gradient, of the straight-line graph is the Hubble constant, which is here assumed to be 15 miles a second per million light-years.

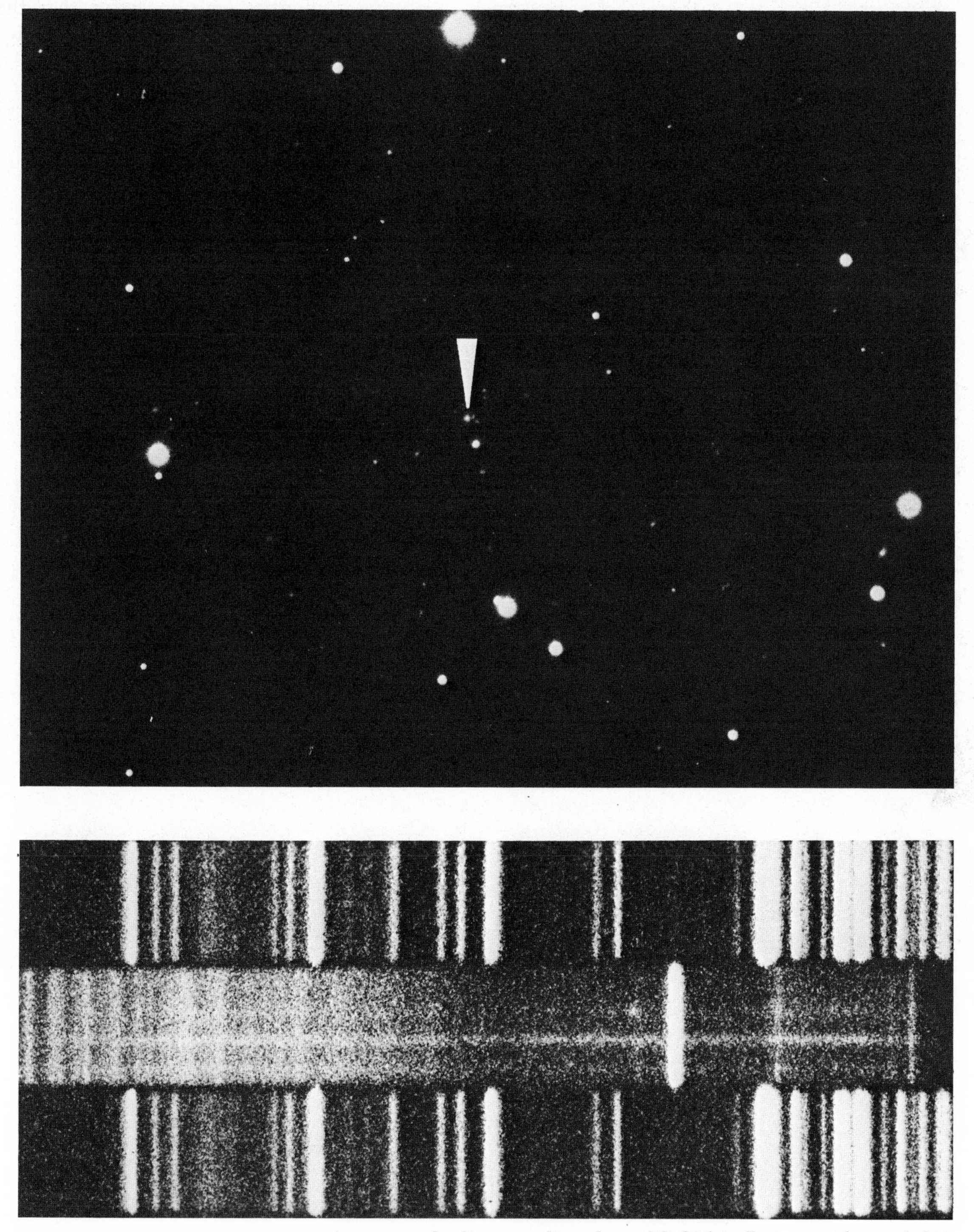

An extremely distant radio galaxy, 3C 295 in Boötes, with a red-shift of 41 percent of the wavelength and a velocity of recession of 76,000 miles a second.

—Mount Wilson and Palomar Observatories

the member galaxies. The fact that the velocity-distance relation seemed to be valid over so vast a volume of space indicated that it was not only a powerful research tool but also a clue to the structure of the universe.

Nor was this all. After 1949, when Humason used ultrarapid spectrographs applied to the 200-inch telescope on Palomar Mountain, Hubble was able to extend his relation to several even more distant clusters, among them one in Hydra with an estimated velocity of 38,000 miles a second, or roughly one fifth the velocity of light. More recently, R. Minkowski at Palomar has found that 3C 295, the most distant galaxy yet photographed, has a recession rate of 76,000 miles a second, or about two fifths the velocity of light. So if we assume that Hubble's law holds indefinitely, the distance of 3C 295 is about 760 million light-years.

Since velocities of this high order are not easy to digest, we are inclined to question the original value of the Hubble constant, and also ask whether the velocity-distance relation is linear out to 3C 295. We may even feel like questioning the validity of the assumption underlying it all, namely, that the observed shifts of the spectral lines are due solely to velocities of recession.

INTERPRETATION OF THE RED-SHIFT There is no question of the reality of the red-shifts. They are beyond dispute, but the velocities based on them are not. The lines usually selected for measurement are the H and K absorption lines of calcium since they are strong in the spectra of most galaxies. In those of very remote galaxies some lines are shifted beyond even the red but can still be detected and measured by using photographic plates sensitized to the infrared. Associated with the red-shift is an overall reduction in the energy of the radiation emitted and an increase in the wavelength of the maximum energy level or intensity. In other words, a galaxy with a large red-shift is appreciably fainter and redder than it would be if its radial velocity were zero. Fortunately the reduction in brightness and the amount of reddening can be calculated from the red-shift and allowed for when trying to establish luminosities and distances.

Some authorities have suggested that the red-shift might be the result of a progressive loss of the energy of light as it travels over the vast distances of intergalactic space. For a photon (a tiny unit of light energy) the product of energy and wavelength is constant. So if the energy were to decrease the wavelength would increase and give rise to a Doppler effect. This hypothesis, however, has had a poor reception, mainly because of the complete lack of observational support from other areas of physics and also because it invokes a new principle of nature, namely, that light loses energy as it gets older.

Light certainly loses energy when it does work in escaping from a stronger to a weaker gravitational field. This brings about an increase in wavelength which shows up as a red-shift. The effect, known as the Einstein gravitational red-shift, is found in the spectra of white dwarfs like Sirius B and 40 Eridani B and also in the spectrum of the sun, but it is most unlikely that it could account for the large shifts in the spectra of galaxies. One would have to suppose that galaxies acquired ever stronger gravitational fields as their distances from our galaxy increased.

Much of the objection to the interpretation of the red-shift as a Doppler effect is perhaps psychological. For most of us velocity is a fairly familiar concept drawn from the ordinary world of everyday experience. In such a world a velocity of the order of 38,000 miles a second is so large as to seem almost incredible. Yet if a galaxy comparable in size to our galaxy moved at this rate it would still take about 500,000 years to pass through a distance equal to its own diameter.

Model Universes

Red-shifts have been observed for too short a time to enable us to say whether they were smaller or larger in the past than they are now. If we assume that the red-shift is a Doppler effect and also that the ratio $\Delta\lambda/\lambda$ is constant for any one galaxy, we have a universe whose rate of expansion is constant. That the universe is expanding was to some extent predicted in

1922 by the Russian mathematician A. A. Friedmann. Particular solutions of certain equations in Einstein's general theory of relativity led Friedmann to formulate two model nonstatic universes. In one, the so-called oscillating model, the universe alternately expands and contracts. At maximum contraction all its material is compressed to a certain maximum density. It then expands to reach a certain minimum density, after which it contracts, expands again, and so on ad infinitum. In the other, the "hyperbolic" model, the universe contracts from a state of infinitely low density to one of a certain maximum density and then expands to an indefinite extent.

THE "BIG-BANG" OF LEMAÎTRE'S THEORY In 1927, Georges Lemaître, a Belgian mathematician, drew attention to yet another "nonstatic" solution of Einstein's equations. According to this all the matter in the universe was once confined to an extremely small space in a state of ultrahigh compression and extreme heat. This giant nucleus or "primeval atom" then exploded and its fragments, in the form of galaxies, were sent flying in all directions.

The question of how long ago the explosion or "big-bang" took place was first answered by Hubble. If space is expanding uniformly the galaxies embedded in it indicate its motion in much the same way as floating straws indicate the currents in a stream. So by working backwards it is possible to determine the epoch T_0 years ago when the galaxies were all bunched together. To obtain T_0 we have only to divide the distance (D) of a galaxy by its velocity (V). When we do this we are in effect finding the reciprocal of the Hubble constant, for $T_0 = D/V = 1/H$. According to Hubble the value of T_0 was 1.8 billion years. This represented the time it would take a galaxy to travel 1 million light-years at a velocity of 100 miles a second.

THE PROBLEM OF THE TIME SCALE But did 1.8 billion years represent the age of the universe? Apparently not. The time scale was disconcertingly short. Measurements of the relative abundances of uranium and other radioactive elements in surface rocks indicated that the age of the earth was at least 2 or 3 billion years. If the earth had this great age it seemed highly probable that the sun was even older, and so presumably were the millions of stars that formed the galaxies.

VALUE OF THE HUBBLE CONSTANT One way out of the difficulty was to adopt the pulsating model and suppose that the galaxy was formed when the radius of the universe was last a minimum. Another was to suppose that the rate of expansion had been slower in the past than now, or alternately, that the rate of evolution of stars and systems of stars had been faster. The difficulty was temporarily resolved in 1952 when Baade announced that the extragalactic distance scale based on the Cepheid period-luminosity relation was faulty. This soon led to an increase of the distance scale by a factor of 2.5 and a corresponding reduction in the value of the Hubble constant. At the time Sandage suggested that the Hubble constant should be 40 miles a second per million light-years, with reciprocal of 5 billion years. Since then the constant has undergone further reductions and it now, according to recent investigations by Sandage, lies in the region of 15 miles a second per million light-years. On the basis of this last value T_0 becomes 12 billion years and the universe acquires a time scale more in keeping with that of the ages of the stars.

If we make the Hubble constant 15 miles a second per million light-years, 3C 295 shoots out to a distance of 5 billion light-years. This, of course, is a purely nominal figure and will have to remain so until we can command a method of distance determination that is at once reliable and independent of measurements of red-shifts. In any case it is based on the formulas $V = c\Delta\lambda/\lambda$ and $D = V/H$ instead of on more complicated relativistic formulas which should always be used when V is larger than about one third the velocity of light. Theory suggests that the Hubble constant is a first approximation and not a constant at all. Many writers overlook this important point and unreservedly make D equal to 12 billion light-years when $V = c$, the velocity of light. This implies that galaxies at this critical distance have zero brightness and are therefore not visible at all, and also that all the galaxies

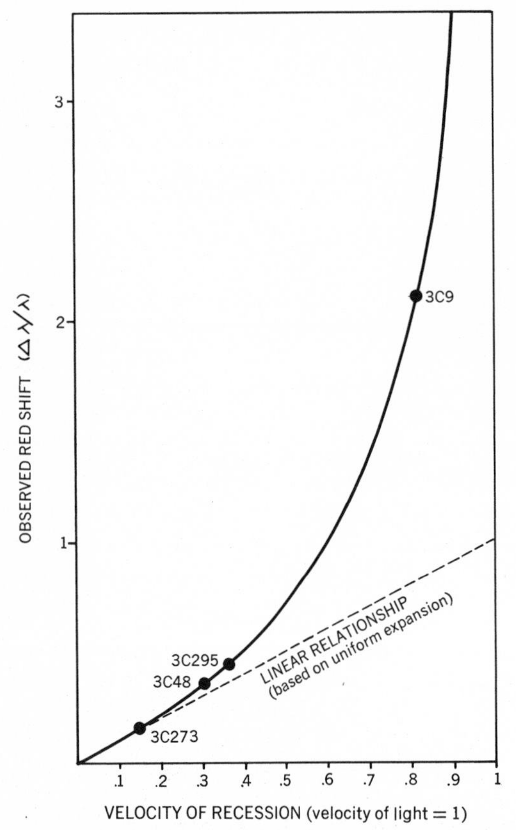

Curve showing how observed red-shift is related to the velocity of recession. 3C 295 is a radio galaxy, 3C 273 and 3C 48 are quasi-stellar radio sources and 3C 9, with a red-shift of 2.1, is the most distant known quasi-stellar radio source. Velocities of recession approach but can never quite reach the velocity of light. Based on a diagram by J. L. Greenstein in *Scientific American*, Dec. 1963

now visible will eventually cross the threshold into invisibility. Yet the relativistic formulas show that the velocity of recession approaches but never quite reaches the velocity of light. A galaxy could not therefore reach a state of invisibility. For very high speeds of nine tenths the velocity of light, its ultraviolet emission would be shifted well into the infrared, but the gamma-ray emission would then be in the visible region.

THE GEOMETRY OF SPACE According to the general theory of relativity, space is not necessarily Euclidean, or infinite and flat (zero curvature). It may be curved in such a way as to be either open and infinite (negative curvature), or closed and finite (positive curvature). In space of zero curvature the volume of a sphere increases in strict proportion to the cube of the radius. In negatively curved, or hyperbolic, space the volume increases faster than the cube of the radius, while in positively curved, or spherical, space it increases slower. One way of determining the geometry of intergalactic space, therefore, is to count the number of galaxies within different volumes of space. If the number increases at a rate faster than the cube of the distance, space is hyperbolic and the universe is open and infinite. If the number increases at a rate slower than the cube of the distance, space is spherical and the universe is closed and finite.

When Hubble carried out this test he found that certain discrepancies in the velocity-distance relation indicated that space had positive curvature and that the universe was therefore finite in volume and contents. The precise value of the curvature could not be ascertained, but he thought that the regions he and Humason had surveyed were already a large fraction of the whole. The work involved correcting for the reduction in apparent brightness due to the red-shift, for without the correction the galaxies apparently thinned out as the distance increased.

Another complication is that the luminosity of a galaxy probably changes with time. Galaxies, like stars, are evolving, so we should expect them to decrease in luminosity with increasing age. If they were formed as a direct result of an explosion 12 billion years ago, they should all be of about the same age. We see them, however, at different times in the past according to their distances from us. Very remote galaxies are seen as they were several billion years ago and could therefore look younger and more luminous than those in our immediate neighborhood. But as it is not known for sure how galaxies evolve we have no way of determining how luminosity changes with increasing age.

Another way of deciding whether the curvature of space is zero, positive, or negative is to find out how the velocity-distance relationship behaves when the distances are very great.

Theory shows that the relationship is approximately linear to a distance of about 1 billion light-years. It then deviates from the linear by predictable amounts according to the type of space. The deviations increase with distance and are greatest for space of positive curvature and least for space of negative curvature. They all have an upward trend in the sense that as the distance increases the velocity progressively exceeds that of strict proportionality. This is equivalent to saying that the rate of expansion steadily decreases with increasing distance: remote clusters of galaxies are also remote in past time and therefore move faster than in direct proportion to their distance. But before the deviations can be submitted to observational tests with reasonable chance of success we need reliable distance determinations.

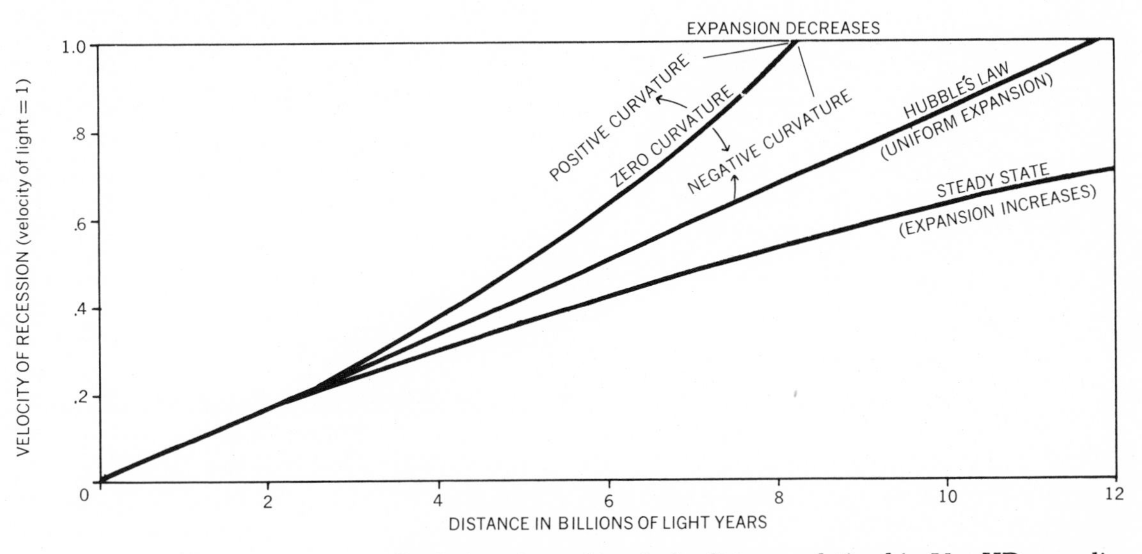

Deviations from the velocity-distance relationship V = HD according to different types of space geometry. The distance scale is based on estimates of apparent brightness and is therefore provisional.

An expanding universe has the important property that the general or large-scale view is the same from each and every galaxy. An observer on 3C 295, for example, would not only see galaxies in all directions but also find that they were all dashing away from him with velocities proportional to their distances. In other words an expanding universe has no absolute center and no absolute boundary or "edge." There are as many centers and edges as there are galaxies. The same argument applies when we consider the location of the "big bang." Our observations indicate that it was located in our galaxy at time T_0, but an observer on another galaxy would place it in his galaxy. In brief, if we accept the "big bang" theory, we have to recognize that the universe had no *absolute* center of origin.

THE STEADY-STATE THEORY Nonstatic, or "evolutionary," models of the universe now have a strong competitor in the "steady-state" model, introduced about 1948 by H. Bondi, T. Gold, and F. Hoyle. According to the latter the universe not only has no center and no edge but also no beginning and no end in time. Its space is Euclidean and therefore infinite, and the rate of recession of its galaxies should steadily increase with increasing distance. Wherever an observer may be located in space and time he always gets the same overall, or large-scale, picture. The average density of matter in the universe as a whole is therefore constant, even though the galaxies are flying away from one another. To keep the density constant, Bondi, Gold, and Hoyle postulated the continuous creation of matter. Somehow hydrogen atoms are newly formed in space. They collect to build new galaxies which, so to speak, fill in the regions of low density brought about by expansion.

The theory of the steady-state universe has been criticized on numerous grounds, particularly by those astronomers who seek quantitative information rather than statements of how the universe should behave. As originally formulated the theory stated that the density of matter on the large scale is constant but did not say what that density should be. It asserted that matter is formed continuously but shed no light on the mode of creation. Since matter is a form of energy and cannot be created out of nothing, some theoreticians have postulated a source of negative energy which they call the C-field. They have also attempted to describe the theory quantitatively and to consider its implications with regard to problems like the origin and distribution of the elements and the formation and evolution of galaxies.

In 1961 M. Ryle of the Cavendish Laboratory in Cambridge, England, made out what appeared to be a strong case against the steady-state theory. He and his associates had succeeded in measuring the positions and intensities of several thousand discrete radio sources. Only a few of these sources had been optically identified, so it was believed that the great majority were extremely distant objects similar to radio galaxies. Ryle then arranged them in a series of ranges in intensity, only to find that the weakest (*i.e.*, the most distant) were about three times more numerous than could be expected if their distribution in space were uniform. This finding could not possibly be reconciled with the steady-state theory if, as Ryle surmised, the sources were extremely remote. On the other hand it fitted the evolutionary picture quite well, for as we look at distant objects (and therefore far back in time) we should expect to find them more numerous than in our immediate vicinity. This interpretation, however, means that the sources decrease in intrinsic power as time advances. Those formed long ago would tend to be intrinsically more intense than those formed recently, so as we look far back in time we should expect to see an excess of intense sources. Yet according to the steady-state theory we should expect just the opposite: a galaxy would be more likely to become a strong radio source as it grew older.

Needless to say the supporters of the steady-state cosmology questioned not only Ryle's inferences but also the accuracy of his data. The validity of the assumptions that the observed intensities were reliable distance indicators, and that the majority of the sources were beyond the reach of giant optical telescopes, came under heavy fire. Doubts were expressed as to the accuracy of the measurements and the statistical adequacy of the sample. It was also suggested that local clusterings of the sources might account for some of the results. These and other considerations have since been examined in some detail, as well as the question of the nature of the sources, their probable distances, and their evolutionary significance, but no general decision between the two theories has been reached.

Quasars

Models of the universe now have to reckon with small but intense sources of light and radio waves known variously as "quasi-stellar radio sources," "quasars," and "QSS's." The first three, discovered in 1962 by T. Matthews and Sandage, were thought to be stars in our galaxy, but further investigations showed that they differed from normal stars in at least three ways. Their spectra consisted of a continuous background crossed by a few strong, broad emission lines at unexpected wavelengths; they emitted abnormally large amounts of ultraviolet radiation; two of them, 3C 48 and 3C 196, were found to vary in brightness. Then, to make interpretation more difficult, Sandage discovered that 3C 48 and 3C 196 were accompanied by faint reddish wisps of nebulosity.

In 1963 M. Schmidt of the Mount Wilson and Palomar Observatories found that the peculiar appearance of the spectrum of 3C 273, a QSS in Virgo, could be described in terms of the Doppler effect. If one assumed a red-shift of 16 percent of the wavelength the enigmatic emission lines corresponded with the familiar lines of hydrogen and doubly ionized oxygen.

An examination of the spectrum of 3C 48 led to a similar conclusion. It looked as though at least two QSS's were not in the Milky Way System at all but out among the radio galaxies.

As is so often the case in science, these discoveries raise more problems than they solve. The red-shift of 3C 273 indicates that it has a velocity of recession of about one seventh that of light and a distance of 2 billion light-years. This in itself is not surprising. What is surprising is the fact that despite its great distance it looks like a star of the thirteenth magnitude. If it were a normal galaxy its magnitude would be about 20. So although it cannot be as large as a normal galaxy, it is about one hundred times as luminous as M31 and about two hundred times as luminous as our galaxy. This brightness, coupled with remarkably high intensities in the ultraviolet and millimeter parts of the spectrum, denotes an overall radiative output far and away larger than that of any of the other intense sources we have considered. A comparison of the optical and radio emissions, together with the fact that the radio waves are partially polarized, shows that much of the radiation is due to the synchrotron process, and this in turn suggests a prodigious explosion. The power of the explosion in M82 has been described as equivalent to a hydrogen bomb 1 million times as massive as the sun, but that of 3C 273, according to Greenstein, is one hundred to one thousand times greater than that.

That 3C 273 is in the throes of mighty changes is shown on photographs taken by Schmidt with the 200-inch telescope. A faint luminous jet extends from the starlike image to a distance of at least 130,000 light-years. This presumably accounts for the curious radio structure discovered by Australian observers with the 210-foot radio telescope at Parkes. As a radio source 3C 273 has two main components separated by about 20 seconds of arc. The stronger, component *A*, is markedly elliptical and coincides with the luminous jet. The weaker, component *B*, consists of two parts which together are associated with the starlike image. Component *B* is also variable in radio intensity and has undergone a marked increase in intensity in the last two to three years. Even more surprising are the big changes in emission at millimeter wavelengths; preliminary observations indicate that the energy from this part of the spectrum can almost equal that from the visual part.

QUASARS THAT FLUCTUATE IN ENERGY OUTPUT Photographs taken over the last eighty years or so show that 3C 273 fluctuates in brightness by about half a magnitude over intervals of about ten to thirteen years. There are also occasional short-lived increases of a magnitude, or more than two-fold surges in brightness. Since material disturbances cannot be propagated across a body with a speed exceeding the velocity of light, most of the fluctuating light must come from a relatively small source. Its diameter cannot exceed the product of the velocity of light and the period of light variation. Since the surges last about a week, the diameter of the superluminous central part of 3C 273 must be less than about seven light-days, or less than fifty times the diameter of the orbit of Neptune. Whether the light changes are related to the changes in intensity at radio wavelengths is not yet established, but there are already indications that this may be the case.

Since 3C 273 lies about 10 degrees south of Virgo A, their radio intensities can be compared quite easily. It is also one of the nearest, if not the nearest object of its kind: the others lie between apparent magnitudes 16 and 19 and are presumably more distant. Their spectra invariably show emission features, which when allowance is made for the red-shifts, correspond to several of those in the spectra of planetary nebulae. Their outer parts must therefore be gaseous, and this, considered along with their exceptionally high luminosities, large outputs of synchrotron radiation, and excesses in the ultraviolet, points to upheavals of an explosive nature. A large percentage of the seventy-four objects so far investigated show variations in brightness, while many, like 3C 273, vary considerably in radio intensity.

QUASARS AND THE DISTANCE SCALE The fact that QSS's have remarkably high luminosities means that they can be observed at greater distances than normal galaxies. One of the most

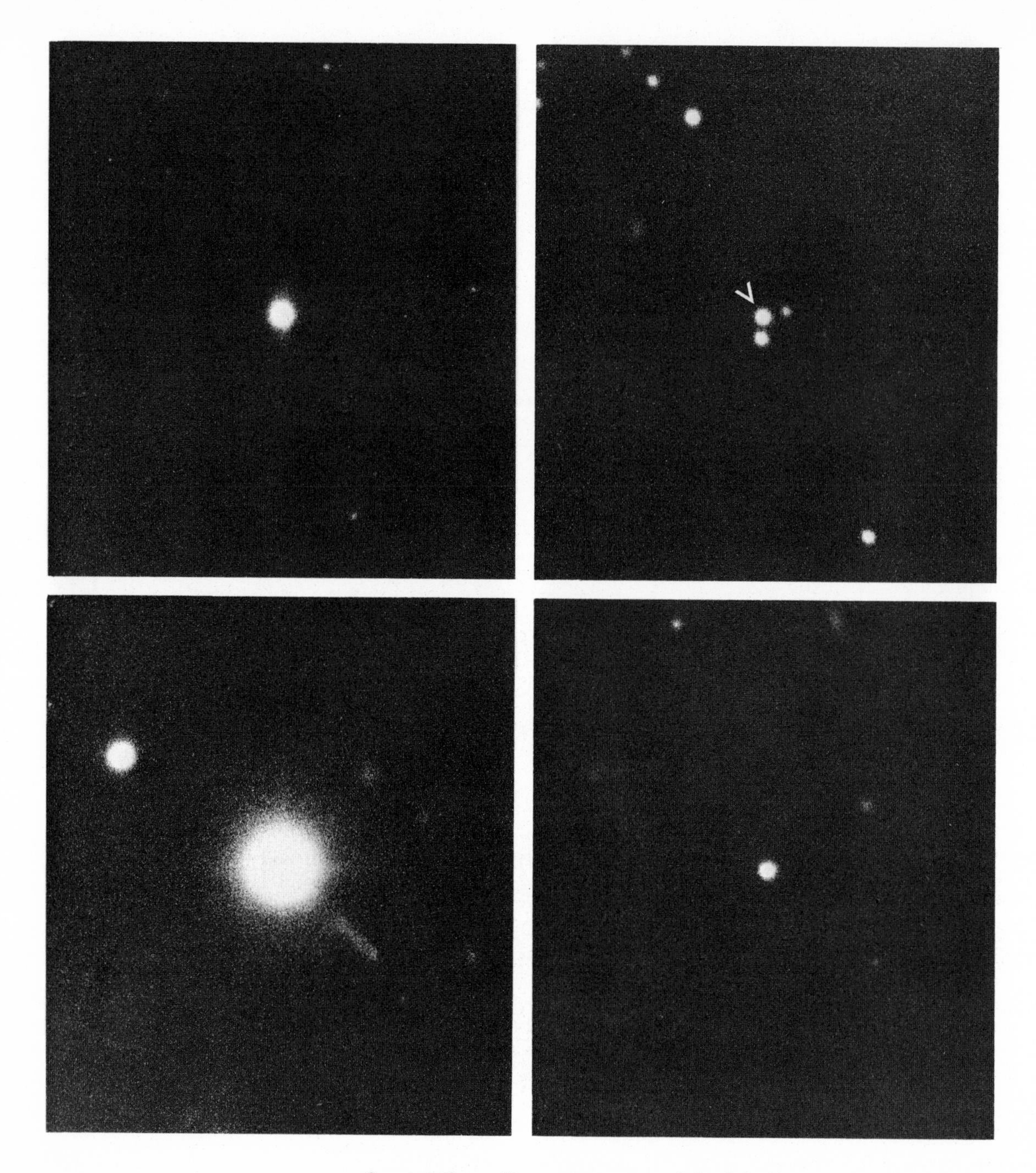

Quasi-stellar radio sources: upper left, 3C 48; upper right, 3C 147; lower left, 3C 273; lower right, 3C 196.
—MOUNT WILSON AND PALOMAR OBSERVATORIES

distant is 3C 9, with a red-shift of 201 percent of the wavelength. This does not mean that its velocity is twice that of light, for as we saw earlier the formulas $V = c\Delta\lambda/\lambda$ and $D = V/H$ are valid only for relatively small cosmological velocities and distances. When we use the corresponding relativistic formulas the velocity of 3C 9 comes out to about four fifths the velocity of light.

Whether QSS's will eventually provide vital information about the structure of the universe will depend largely on their abundance. At the time of writing over a hundred have been detected, but there may well be large numbers on and beyond the present thresholds of observation. The properties of those studied have led to much speculation and discussion, but a clearer picture of their significance will not emerge until much more is known about their distances, physical dimensions and variations in intensity. Not all astronomers are prepared to accept the Doppler interpretation of the red-shifts, and some of those who do consider that many of the fainter QSS's are fast-moving objects near and perhaps even in our own galaxy. However, 3C 295 is almost certainly extragalactic, for J. A. Koehler of the Arecibo Radio Observatory, Puerto Rico, has obtained evidence that strongly suggests that it lies *beyond* the Virgo cluster. Its radio spectrum contains an absorption line due to neutral hydrogen, but shifted away from the 21-centimeter region by an amount that shows that the gas cloud responsible has a velocity of recession of about 700 miles a second. Since this is roughly the velocity of the Virgo cluster, Koehler concluded that the absorption was occurring in the cluster itself.

QUASI-STELLAR GALAXIES Much more likely candidates for resolving cosmological issues are intensely luminous starlike objects which look like QSS's but whose radio intensities are relatively low. Their discovery was announced by Sandage in June, 1965. They had hitherto been regarded as hot, blue stars in the galaxy, but Sandage found that they tended to avoid the region of the Milky Way and looked uncommonly like QSS's. He and Schmidt then found that the lines in the spectra of some of them had large red shifts. Preliminary observations indicate that they are extragalactic, highly luminous, and very plentiful. Sandage has labeled them "quasi-stellar galaxies" or QSG's, and considers that they represent a major new constitutent of the universe.

THE FORMATION OF ELEMENTS If, as Lemaître suggested, the universe started from a single giant atom or nucleus of exceptionally great density, all the first-formed elements would have been extremely heavy. How, then, were the lighter elements formed, and how can we account for their present high relative abundances? Since Lemaître's theory shed no light on this problem, G. Gamow and others proposed that the original material consisted of hot nuclear gas. Immediately after the explosion the neutrons composing the gas decayed into protons and electrons, and in a rapid series of thermonuclear processes all three combined to form the complete range of elements. In other words, the elements, heavy as well as light, were all "cooked" during the first half-hour or so of the explosion.

This is a vastly different picture from that provided by the modern theory of stellar evolution. As we have seen, the basic material involved in the birth of stars, and therefore of galaxies, would appear to be hydrogen. This is then synthesized into helium and other elements in the hot interiors of stars, and these heavier elements are eventually injected into the interstellar medium as the stars explode, throw off their surface layers, or lose mass in other ways not yet fully understood. So, if a new generation of stars condenses from the medium, its members are assured of a percentage of heavier elements as their birthright.

ORIGIN OF THE GALAXIES It would therefore seem that the galaxies did not arise from complex conglomerations of atoms of different elements but rather from condensations in thin expanding hydrogen gas. The gravitational force in a condensation must have been sufficient to overcome the general velocity of the expanding gas, thus enabling it to contract and acquire through rotation a more regular shape. Local condensations within the mass then contract separately to form stars and begin the work of building up all the different elements. Yet while this may seem straightforward and plausible, its details are extremely involved and puzzling. Theories of galaxy formation based on condensations and perturbations in a diffuse medium have all run into serious difficulties.

An intriguing question is whether a diffuse medium pervades the space between the galaxies. Material is undoubtedly being expelled from exploding and interacting galaxies, and also perhaps from many QSS's. Koehler's observation of the 21-centimeter line of neutral hydrogen in the spectrum of 3C 295 indicates that this gas is present in the Virgo cluster. Also present in the cluster is NGC 4189, a spiral galaxy with a faint dark patch, which according to C. T. Kowal could be an intervening dust cloud. IC 3258, another member of the same cluster, has a similar dark patch; so also has an unnamed galaxy in Pisces. F. Zwicky finds that the number of faint clusters of galaxies tends to decrease in regions of the sky covered by nearby rich clusters. In his opinion intergalactic matter not only accounts for certain observed irregularities in the distribution of clusters of galaxies, but also, by its dimming and reddening effect, effectively reduces the total number of galaxies within reach of the 200-inch telescope.

Some astronomers think that the clue to the origin of galaxies may lie in the QSS's. They argue that since these objects release energy at a fantastic rate they are presumably the direct results of explosions and are therefore comparatively short-lived. Preliminary calculations give them ages of the order of 100,000 years to 1 million years, while QSG's, according to Sandage, last for about 500 million years. Are QSS's and QSG's galaxies in the initial stages of their formation? If they are they could not have been born soon after the "big-bang" of the evolutionary universe but are being born more or less continuously. On the other hand M. S. Longair at Cambridge, England, has recently constructed a model based on counts of QSS's which puts the time of the formation of these objects at about 200 million years after the "big bang."

ORIGIN OF THE ENERGY OF QUASARS The energy output of a QSS is difficult to account for in terms of thermonuclear reactions. The sudden release of the energy of 100 million stars might do the trick, but this would require that they all be confined to a small volume of space and then simultaneously become supernovae. A promising alternative, put forward by Hoyle and W. A. Fowler, is gravitational energy. This arises when a body suddenly collapses, the amount released being proportional to the square of the mass involved. Thus if a whole galaxy were suddenly to collapse or "implode" into a volume one thousand light-years across, the energy released would be roughly that of a QSS. Such a process could feasibly occur in a great mass of gas in which star formation, instead of running its usual course, is partly or completely inhibited. In other words, QSS's may be imploding protogalaxies.

The concept of gravitational energy plays an important part in current theoretical models of QSS's. There is a growing body of opinion that these objects have at their centers a massive superluminous body in a state of gravitational collapse. Around this relatively small core is a spherical shell of hot material several light-years in radius. This gives rise to the continuous radiation. Then comes a cooler second shell containing hot ionized gas. This is a hundred or more light-years across and gives rise to the strong emission lines in the optical spectrum. The radio region is larger still, being of the order of one thousand light-years across, and with its magnetic field and streams of fast-moving electrified particles, produces synchrotron radiation and cosmic rays. Why the structure should dispense its energy in cycles and how the cycles are maintained are mysteries. Theory alone cannot be expected to settle these and the many other outstanding problems in cosmology. Their solution rests on the data of observation, and the data, by the very nature of things, is inadequate. When they are solved a whole crop of new problems will arise, but this is as it should be if our knowledge, like the universe, is to remain in a state of expansion. Meanwhile we must constantly revise our models and theories so that they fit observation, and resign ourselves to the possibility that cosmology will continue to be a controversial subject for a long time to come.

GLOSSARY

ABERRATION OF LIGHT The apparent displacement in the position of a star owing to the effect of the earth's velocity in its orbit combined with the velocity of light. The displacement is greatest for stars lying in a direction at right angles to the direction of the earth's motion.

ABSORPTION LINES Dark lines in a continuous spectrum produced by the absorption of light of certain wavelengths by a layer or layers of gas cooler than the source of light.

ALTITUDE The angular distance of a body above the horizon.

ANGULAR VELOCITY The rate of change of the angle swept out by the line drawn from a moving body to the center about which it moves.

ANTAPEX, SOLAR The point on the celestial sphere away from which the sun is moving.

APERTURE The effective opening or working diameter of an optical system. In the case of a telescope it is usually the diameter of the object-glass or primary mirror.

APEX, SOLAR The point on the celestial sphere toward which the sun is moving.

ARIES, FIRST POINT OF The point on the celestial sphere reached by the sun at the time of the vernal equinox, about March 21. It was formerly in the constellation of Aries, but owing to the precession of the equinoxes it now lies in Pisces.

ASTRONOMICAL UNIT The mean or average distance of the earth from the sun—roughly 93 million miles.

AURORA Colored lights emitted in the upper atmosphere at heights roughly between 50 and 600 miles. They are caused mainly by electrified particles from the sun.

BINARY A system of two stars, each of which revolves about the common mass-center of the system.

BLACK BODY An ideal or imaginary body whose total emissive power is proportional to the fourth power of its absolute temperature. A close practical approximation to this ideal is a hollow enclosure whose inner wall has a uniform temperature and a small hole through which the radiation can escape.

CELESTIAL EQUATOR The projection of the earth's equator on the celestial sphere.

CELESTIAL POLES The two points where the earth's axis, continued into space, meets the celestial sphere.

CELESTIAL SPHERE The imaginary sphere on which the stars appear to be fixed. It is centered on the observer and has an infinitely large radius.

CEPHEIDS Variable stars whose pattern of light behavior is similar to that of Delta Cephei. Their period of light variation can be anything from about a day to over 50 days and in all cases is related to the luminosity.

CHROMOSPHERE The lower layer of the sun's atmosphere. It consists mainly of hydrogen and is red in color.

CIRCUMPOLAR STAR A star that does not rise or set.

CLUSTER, STAR A group of stars generically associated and having a common motion through space.

CLUSTER VARIABLE A variable star, first detected in globular clusters, with a period of light variation less than about a day. Its pattern of light behavior is similar to that of RR Lyrae.

COLOR INDEX The difference between the photographic and visual magnitudes of a star.

CONSTELLATIONS The groups or divisions in which the stars are arranged for purposes of identification. The practice arose in ancient times from the desire to picture various objects among the stars. The pictures are known as constellation figures.

CORONA The outer part of the sun's atmosphere that appears as a crown of pearly white light around the new moon when the sun is totally eclipsed.

CORONAGRAPH A telescope designed to photograph the sun's corona and prominences in broad daylight.

COSMIC RAYS Fast-moving electrified particles which reach the earth from outer space.

COSMOGONY The study of the origin of the physical universe or any of its constituent parts.

COSMOLOGY The study of the nature and structure of the physical universe.

DECLINATION The angular distance of a celestial body north (+) or south (−) of the celestial equator. It is measured along a great circle passing through the celestial poles and corresponds to latitude on the earth.

DIFFRACTION The slight deflection and dispersion of light when it passes the edge of an obstacle. It occurs at the edge of a mirror or lens and can give rise to a series of circular fringes surrounding the images of stars.

DISPERSION OF LIGHT The separation of white light into its constituent colors.

DOPPLER EFFECT A shift in the spectrum lines of a source moving toward or away from an observer. The shift is measured relative to the corresponding spectrum lines of a laboratory source.

DOUBLE STAR Two stars which appear close together in the sky either because they happen to be in almost the same direction (OPTICAL DOUBLE) or because they are physically associated (BINARY).

DURCHMUSTERUNG A star catalog that lists the positions and some of the properties of all stars down to a certain magnitude over a large part of the sky.

DWARF STARS Stars on the main sequence. They are low in luminosity and small in size compared with giant stars.

ECLIPSE The passage of one celestial body through the shadow of another. Thus an eclipse of the sun occurs when the earth passes through the shadow of the moon.

ECLIPSING BINARY A system of two stars whose components periodically occult, or cut off, part or all of the light of each other.

ECLIPTIC The apparent path of the sun relative to the background of stars. It is also the projection of the earth's orbit on the celestial sphere.

ELECTROMAGNETIC RADIATION Energy in the form of rays or waves, *e.g.*, gamma rays, X rays, ultraviolet, visible light, infrared, radio waves.

ELECTRON An elementary particle of unit charge of negative electricity.

ELEMENT In chemistry a substance reduced to its simplest possible form.

ELLIPSE A geometrical curve drawn about two fixed points, called foci. The sum of the distances between any one point on the curve and the foci is constant and equal to the greatest diameter of the ellipse.

EMISSION LINES Bright lines in a spectrum produced by the emission of light of certain wavelengths by a glowing gas.

EPOCH A date of reference employed in astronomical calculations. When used in connection with star charts and catalogs it is the date at which the co-ordinates given will be strictly correct.

EQUINOCTIAL POINTS The two points where the ecliptic intersects the celestial equator. The sun passes through these points at the equinoxes.

EQUINOXES The two times of the year when daytime and night each equal 12 hours—about March 21 (vernal equinox) and about September 22 (autumnal equinox).

EXTRAGALACTIC Outside our galaxy, or Milky Way system.

EYEPIECE The lens or combination of lenses placed at the eye end of a telescope. Its function is to magnify the image formed by the object-glass or primary mirror.

FACULAE Bright spots and streaks on the sun's photosphere.

FLARE, SOLAR A sudden and brief brilliant eruption on the sun with associated emissions of radio waves, ultraviolet radiation, and streams of electrified particles.

FLARE STAR A star subject to sudden and often quite considerable increases in luminosity.

FLOCCULI Bright and dark patches on the surface of the sun; best shown on spectroheliograms.

FLUORESCENCE The process in which radiation of short wavelength is absorbed and re-emitted at longer wavelengths.

GALACTIC EQUATOR A great circle about one degree north of the center line of the Milky Way.

GALACTIC POLES Two opposite points on the celestial sphere 90 degrees from the galactic equator.

GALAXIES Large gravitational systems of stars.

GALAXY, THE The Milky Way System, or great system of stars, gas, and dust to which the sun belongs.

GAMMA RAYS Electromagnetic radiation of extremely short wavelengths.

GAUSS The unit in which the intensity of a magnetic field is measured. When a single loop of wire of radius 0.628 centimeter carries a current of 1 ampere the intensity of the magnetic field at its center is 1 gauss. The intensity of the magnetic field near the earth's surface has a range of about 0.3 to 0.6 gauss.

GIANT STARS Stars more luminous and therefore larger than stars on the main sequence (DWARF STARS) of the same spectral type.

GRAVITATION The tendency of all bodies in the physical universe to attract one another with a force directly proportional to the product of their masses and inversely proportional to the square of their distance apart.

GRAVITY, CENTER OF The point where the entire mass of a body or system of bodies may be considered to be concentrated and to act. Also known as the mass-center.

GREAT CIRCLE A circle on a sphere which divides the sphere into two equal parts or hemispheres. The shortest distance between two points on the surface of a sphere is an arc of a great circle.

HELIUM A light element discovered through its emission lines in the spectrum of the sun's atmosphere during the total solar eclipse of 1868. Its presence on the earth was not discovered until 1895.

HORIZON The great circle formed by the intersection of the tangent plane to the surface of the earth at the place of observation with the celestial sphere.

H-R DIAGRAM The diagram formed when the absolute magnitudes of stars are plotted against their spectral classes or temperatures. It is named after E. Hertzsprung and H. N. Russell, pioneers in the study of stellar types, and was first introduced by Russell in 1913.

HUBBLE CONSTANT The factor of proportionality connecting the velocities of recession of the galaxies with their distances. Recent work suggests a value of about 15 miles a second per million light-years.

HYDROGEN-ALPHA A red-colored line in the spectrum of hydrogen.

HYDROGEN I AND II The hydrogen in interstellar space appears to be neutral atomic hydrogen (Hydrogen I) or ionized hydrogen (Hydrogen II).

INFRARED A band of electromagnetic radiation between the red end of the spectrum and the very short radio waves. It has important heating properties.

INTERFEROMETER A device used in optical and radio astronomy for measuring extremely small angles such as the apparent diameters of stars and areas of radio emission.

IONIZATION The process in which an atom loses one or more of its electrons. Since an electron is a unit charge of negative electricity, the atom becomes positively charged or ionized.

LIGHT, VELOCITY OF 186,282 miles per second in a vacuum.

LIGHT-YEAR The distance which light travels in a year. It is approximately equal to 6 million million miles, or 63,290 astronomical units.

LIMB The circular edge of the sun, moon, and any planet as seen in projection on the sky background from the earth or any other specified viewpoint.

LOCAL GROUP The cluster of galaxies which includes, among others, our galaxy, the Clouds of Magellan, M31 in Andromeda, and M33 in Triangulum.

LUMINESCENCE The process in which radiation of short wavelengths is absorbed and re-emitted at longer wavelengths. It is also known as fluorescence.

LUMINOSITY The intrinsic brightness or light output of a star as distinct from its apparent brightness.

MAGNETOGRAM A photographic map of the sun's general magnetic field as obtained with a magnetograph.

MAGNETOGRAPH An instrument devised in 1952 by H. W. Babcock and H. D. Babcock for mapping the sun's general magnetic field.

MAGNETOSPHERE A region of "radiation" surrounding the earth with its greatest extension in the plane of the earth's magnetic equator. The "radiation" consists mainly of electrons and protons temporarily trapped in the earth's magnetic field.

MAGNITUDE, ABSOLUTE The magnitude which a star would have if it were at a distance of 10 parsecs from the earth. This distance corresponds to a parallax of 0.1 second of arc.

MAGNITUDE, APPARENT A number representing the relative brightness of a star or starlike object. A star of the first magnitude is said to be 100 times brighter than one of the sixth magnitude. A difference of 1 in magnitude corresponds to a ratio in brightness of $100^{1/5}$, or 2.512.

MAIN SEQUENCE If the stars near the sun are arranged in order of increasing luminosity, most of them form a sequence of increasing mass, surface temperature, and size. These stars are referred to as main-sequence or dwarf stars.

MASS The quantity of matter in a body.

MESSIER NUMBER The number of an object listed in the catalog of 103 nebulae and star clusters prepared by C. Messier in 1784. For brevity an object in the catalog is referred to as M followed by the catalog number.

MOLECULE The smallest quantity of a substance that can exist without losing its chemical identity.

MOMENTUM The product of the mass and velocity of a body.

MONOCHROMATOR An optical filter designed to transmit monochromatic light, or light of a single wavelength.

NADIR The point on the celestial sphere directly opposite to the zenith.

NEBULA A term once applied to any hazy patch of light among the stars but now restricted to a cloud of gas and dust.

NGC NUMBER The number of an object listed in the New General Catalogue of nebulae and star clusters prepared by J. Dreyer and issued in 1888 by the Royal Astronomical Society of London.

NOVA A star which undergoes a sudden and considerable increase in brightness. Changes in its spectrum indicate that it is expelling a substantial part of its atmosphere.

OPTICAL DOUBLE Two stars at different distances from the earth that appear to be close together because they happen to be in almost the same direction.

PARALLAX, ANNUAL The apparent change in the position of a star due to the earth's orbital motion about the sun. Its measure is the angle subtended at the star by the radius of the earth's orbit.

PARSEC A distance corresponding to an annual parallax of 1.0 second of arc. It is roughly equal to 3.26 light-years.

PHOTOELECTRIC A term used to describe the conversion of light energy into electrical energy.

PHOTOMETER, STELLAR An instrument used for comparing the brightness of a star with that of a standard point source of light.

PHOTOSPHERE The "light-sphere," or apparent surface of the sun.

POLARIZATION The process in which light waves are made to vibrate in a particular direction.

PRECESSION OF THE EQUINOXES The slow movement of the equinoctial points along the ecliptic in an east-to-west direction. It is also the motion of a celestial pole about the corresponding pole of the ecliptic.

PROMINENCES Clouds, jets, and streamers of hot gases in the sun's lower atmosphere.

PROPER MOTION The apparent angular movement of a star per year or per century due to its actual motion in space.

PROTON The nucleus of the hydrogen atom.

PROTOSTAR The probable early form of the material that became a star. It is measured directly by means of the DOPPLER EFFECT.

RADIAL VELOCITY The velocity of a body directly toward or away from an observer.

RADIATION Electromagnetic energy in the form of rays or waves, *e.g.*, gamma rays, X rays, light, heat, radio waves. The term is also applied to streams of electrified particles such as electrons and protons. Radiation of this second kind is called particle radiation.

RED-SHIFT A shift toward the red of the lines in the spectrum of a luminous source. Red-shifts are found in the spectra of galaxies and are interpreted as Doppler effects due to velocities of recession.

RIGHT ASCENSION The angular distance of a body from the vernal equinox, or First Point of Aries, measured along the celestial equator.

SCHMIDT TELESCOPE A telescope in which the reflecting surface of the main, or primary, mirror is part of a sphere, and the main optical aberrations are corrected by a thin glass plate mounted before the mirror.

SOLAR WIND Streams of electrified particles which travel outward from the sun.

SPECTROGRAM A photograph of a spectrum.

SPECTROGRAPH An instrument for obtaining spectrograms.

SPECTROHELIOGRAPH An instrument for photographing the sun in the light of one particular wavelength.

SPECTROSCOPE An instrument for seeing spectra.

SPECTROSCOPIC BINARY A star known to be a binary by the regular doubling of the lines in its spectrum. The doubling, a Doppler effect, is produced by the regular to-and-fro motions of the two stars as they travel about the mass center of the system.

SPECTRUM The rainbow-colored band of light formed when white light is dispersed by a prism or diffraction grating.

SPICULES Geyserlike columns of hot gas in the sun's chromosphere.

SUPERGIANT STARS Stars of high luminosity and immense size compared with the sun and other dwarf, or main-sequence, stars.

SUPERNOVA An exceptionally bright nova or "temporary star."

SYNCHROTRON RADIATION Radiation emitted by extremely fast-moving electrified particles when they spiral along the lines of force in a magnetic field.

THERMONUCLEAR REACTION A high-temperature process in which one element is transmuted into another.

ULTRAVIOLET That part of the electromagnetic spectrum between the violet and X rays. It has important chemical effects.

WAVELENGTH The length of a wave of radiation measured from one crest to the next. Also, the distance radiation travels during one complete vibration of the source.

WATT A unit of electrical power or rate of work.

WHITE DWARF Very small stars of low luminosity, high surface temperature and very high density.

X RAYS Electromagnetic radiation beyond the ultraviolet.

ZEEMAN EFFECT The splitting of spectrum lines into two or more components when a source of light is in a strong magnetic field. It was discovered by P. Zeeman in 1896.

ZENITH The point on the celestial sphere directly above the observer.

ZODIAC The band or belt on the celestial sphere in which the sun, moon and planets move. It extends about 8 degrees on either side of the ecliptic. Of the regular planets only Pluto ever is seen outside of it.

ZODIACAL LIGHT A faint cone of light which extends away from the sun and is sometimes seen after sunset or before sunrise on clear, moonless nights.

BIBLIOGRAPHY

Abetti, G., and Hack, M. *Nebulae and Galaxies.* New York: Thomas Y. Crowell, 1964.

Aller, L. H. *Gaseous Nebulae.* London: Chapman and Hall, 1965.

———. *Astrophysics: The Atmospheres of the Sun and Stars.* New York: Ronald Press, 1963.

——— and McLaughlin, D. B. (eds.). *Stellar Structure.* Chicago: University of Chicago Press, 1965.

Baade, W. *Evolution of Stars and Galaxies.* Cambridge: Harvard University Press, 1963.

Baker, R. H. *Astronomy.* Princeton: Van Nostrand, 1965.

Blaauw, W., and Schmidt, M. *Galactic Structure.* Chicago: University of Chicago Press, 1965.

Bok, B. J. and P. F. *The Milky Way.* Cambridge: Harvard University Press, 1957.

Bondi, H. *Relativity and Common Sense.* New York: Doubleday, 1954.

Bray, R. J., and Loughhead, R. E. *Sunspots.* New York: John Wiley, 1964.

Gill, T. P. *The Doppler Effect.* New York: Academic Press, 1965.

Goldberg, L. (ed.). *Annual Review of Astronomy and Astrophysics.* Vols. 3, 4. Palo Alto: Annual Reviews, 1965, 1966.

Greenstein, J. L. *Stellar Atmospheres.* Chicago: University of Chicago Press, 1960.

Hoskin, M. A. *William Herschel and the Construction of the Heavens.* New York: W. W. Norton, 1963.

Hoyle, F. *Galaxies, Nuclei, and Quasars.* New York: Harper and Row, 1965.

King, H. C. *Exploration of the Universe.* New York: New American Library, 1964.

Kopal, Z. *Close Binary Systems.* New York: John Wiley, 1959.

——— (ed.). *Advances in Astronomy and Astrophysics.* Vol. 3. New York: Academic Press, 1965.

Menzel, D. H. *Our Sun.* Cambridge: Harvard University Press, 1959.

———. *A Field Guide to the Stars and Planets.* Boston: Houghton Mifflin, 1964.

Page, T. and L. W. *The Origin of the Solar System.* New York: Macmillan, 1966.

Singh, J. *Great Ideas and Theories of Modern Cosmology.* London: Constable, 1961.

Smith, H. J., and E. van P. *Solar Flares.* New York: Macmillan, 1963.

Steinberg, J. L., and Lequeux, J. *Radio Astronomy.* New York: McGraw Hill, 1963.

Struve, O. *Stellar Evolution.* Princeton: Princeton University Press, 1950.

INDEX

Abell, G. O., 133
aberration of light, 21
absolute luminosity, 54
absolute magnitude, 41–42, 45, 51, 55, 56
absorption:
  intergalactic, 139, 153–154
  interstellar, 108–109, 112–113
Acrux (Alpha Crucis), 51, 54, 64, 65
Adams, W. S., 49, 51, 57
AE Aquarii, 70
AE Aurigae, 98–99
AG Pegasi, 84, 85
Aitken, R. G., 63
Alcor, 64
Alcyone, 15, 104
Aldebaran, 6, 51, 53, 61, 102
Algol, (Beta Persei), 17, 65, 66
Algol A, 66
Algol B, 66
Algol C, 66
Alnitak (Zeta Orionis), 97
Alpha Capricorni, 62
Alpha Centauri, 44, 47, 50, 51, 54, 63
Alpha Ursae Majoris, 102
Altair, 44, 51, 56
Ambarzumian, V. A., 98
Anderson, T. D., 72
Antares (Alpha Scorpii), 17, 51, 53, 61, 65, 83
Apian, Peter, 4
Apus, 18
Aquarius, 13
Aquila (Eagle), 111
Ara, 111
Aratus, 14, 17
Arcturus, 6, 14, 17, 51, 53
Argelander, F. W. A., 22
Argo (the Ship), 111, 121
Aries, 13, 19
Aristarchus of Samos, 3
Aristotle, 2, 4
associations, stellar, 97–98, 102
Atlante Farnesiano, 15
Atlas (star), 104
atlases, star, 17–18
Auriga, 98, 111
AX Persei, 84

Baade, Walther, 114, 128, 129, 139, 147
Babcock, H. D., 30
Babcock, H. W., 30
Bailey, S. I., 80
Bappu, M. K. V., 52
Barnard, E. E., 47, 92
Barnard 72, 93
Barnard's star, 46, 47, 50
Bayer, Johann, 17, 18, 20
Be stars, 75
Becker, W., 117
Bellatrix, 17
Berge, G., 120
Bessel, F. W., 7–8, 21, 46, 47, 102
Beta Canis Majoris, 80
Beta Centauri, 51
Beta Crucis, 51, 54
Beta Lyrae, 68, 72
Betelgeuse, 17, 51, 52, 53, 61, 83
Bethe, H., 59
BF Cygni, 84
"big-bang" theory, 147, 149
Big Dipper (Ursa Major), 2, 14, 64, 102
binaries (double stars), 22, 45–49, 50, 51, 55, 61, 62–69, 70, 73, 76
  components, evolution of, 76
  contact, 69
  eclipsing, 63–69, 71, 73, 85
  general properties, 63
  origin, 75
  relative orbits, 62–63
  spectroscopic, 52, 63–65, 70, 74, 76, 96, 114
  visual, frequency of, 63
Bjerknes, Vilhelm, 30
Blackwell, D. E., 36
Bok, B. J., 95, 117, 119
bolometric luminosity, 54
Bondi, H., 149
Boötes (Wagoner), 14, 145
Boss, Lewis, 102, 103, 113
Bowen, I., 87, 97
Bradley, J., 21, 65
Brahe, Tycho, 5, 6, 17, 19–20, 89
BW Vulpeculae, 80

Cancer, 13, 106, 107
Canes Venatici, 110, 115
Canis Major (Big Dog), 111
Canopus, 17, 51, 52, 53
Capella, 17, 41, 51, 53, 64
Capricornus (Sea Goat), 13
carbon cycle, 39
Carina-Cygnus Arm, 117–118
Cassiopeia, 5, 111
Cassiopeia A, 91
Castor (Alpha Geminorum), 17, 63, 64, 65
catalogs, star, 15, 19, 20–22
Centaurus, 97, 102, 111, 121
Centaurus A, 139, 142
Cepheids, 9, 10, 77–79, 81, 122, 123, 126, 127, 132
  period-luminosity relation, 79, 128
  Population I, 81, 124, 128
  Population II, 58, 81, 110, 115, 128
C-field, 150
Chamaeleon, 18
Chandrasekhar, S., 61

charts, star, 19, 22
Chelae, 13
Chi Cygni, 83
chromosphere, 31–32
CI Cygni, 84
Cicero, 14
circle, galactic, 111
classification, stellar, 51–58
Clerke, Agnes M, 127
Clouds of Magellan, 10, 78, 79, 121–126, 131, 134
clusters, galactic, 102–110
  evolutionary trends in, 107–108
  galaxies, 132–133, 144, 153, 154
  globular, 9, 10, 79–80, 109–110, 115, 122, 124
  M 31 and, 127
Coalsack, 92, 93, 95, 108, 117
color index, 42–43
Coma Berenices, 106, 111, 115, 133, 140
constellation figures, 13, 17–18
constellations, 13–14, 15, 17
  Shapley's, 125
convective envelope, 39
Copernicus, Nicolaus, 2, 3, 4, 6, 21, 24
Corona Australis, 98
Corona Borealis (Northern Crown), 133, 144
corona of sun, 35, 36–39
coronagraphs, 32, 33, 38
cosmic rays, 36
Crab Nebula (M 1; NGC 1952), 74, 89–90, 99, 141
Crawford, J. A., 70, 71
Critchfield, C., 59
Cygnus (Swan), 93, 111
Cygnus A, 138, 139, 142
Cysat, J. B., 95

Davis, L., Jr., 119
declination of stars, 18–19, 20, 21
Delta Cephei, 77, 81, 123
Delta Scuti, 80
Deneb, 51, 52
Deutsch, A. J., 65, 85
diffraction, 5
Digges, Leonard, 5
Digges, Thomas, 4
disk, galactic, 115, 116, 120
Doppler effect, 26, 41, 63, 100, 143, 146, 150
Dorado, 18
Doradus, 121
Double Cluster, 107, 108, 115
Draco (dragon), 86, 127
Dreyer, J. L., 87
Dürer, Albrecht, 18, 19
dust, interstellar, 8, 10, 59, 115
dwarf sequence, 54

ecliptic, 3, 13, 15, 16
Eddington, A., 54, 77, 113
Eggen, Olin J., 60, 105
Einstein, Albert, 39, 49, 147
Einstein gravitational red-shift, 146
Electra, 104
elements, formation of, 153
Epsilon, Aurigae, 66–67
Epsilon Eridani, 50
Epsilon Indi, 50
Epsilon Lyrae, 62, 72
Epsilon Virginis, 17

equator:
  celestial, 13, 16, 18
  solar, 27
equinox:
  autumnal, 13
  vernal, 13, 14
equinoxes, precession of, 13–16, 19
Eratosthenes, 17
Eta Aquilae, 77
Eta Carinae, 94, 95, 100
Eta Geminorum, 61, 65, 83, 85
Eta Ursae Majoris, 102
Eudoxus, 14
evolution, stellar, 50, 58–61, 75, 85

Fabricius, David, 83
faculae, 23
55 Cygni, 93
53 Arietis, 98
52 Cygni, 90, 91
Fish's Mouth, 96
Flamsteed, J., 20
flares, solar, 33, 34–36, 38
Fleming, Mrs. W. P., 80, 122
flocculi:
  calcium, 32
  dark, 32, 33
fluorescence, 87
Fomalhaut, 17, 51, 61
Fornax, 130
40 Eridani B, 49, 146
47 Tucani, 78, 109, 121
Foucault, Léon, 3
Fowler, W. A., 154
Fraunhofer lines of solar spectrum, 32
Friedmann, A. A., 147

galactic coordinate system, 111
galaxies:
  classification, 134–139
  dwarf, 130
  exploding, 140–142, 151
  interacting, 138–139
  neighboring, 121–131
  origin, 153–154
  peculiar, 139
  quasi-stellar, 153, 154
  radial velocities of, 143, 144–148
  radio, 139–142
  universe of, 132–142
Galileo, 5, 6, 23, 24, 25
Gamma Cygni, 94
Gamma Draconis, 20
Gamma Leonis, 63
Gamma Pegasi, 80
gamma rays, 39
Gamma Virginis, 62–63
Gamow, G., 153
Gemini, 13, 111
geocentric theory, 5
Gill, David, 22
globes, star, 14–15, 20
globules, 95
Gold, T., 149
Goodricke, John, 65, 68, 77
granules, solar, 23, 24, 39
gravitation, law of, 3, 62

Great Bear (Wagon), 14
Great Galaxy in Andromeda (M 31), 8, 10, 11, 116, 126–130, 132, 135, 138, 139, 143
Great Nebula in Orion, 8, 86, 94, 95, 96–97, 98, 99
Greenstein, J., 60, 105, 119, 148, 151
Groombridge 1618, 22
Grus (Crane), 18, 111
Gyllenberg, W., 56

Hack, Miss M., 67
Hadley, George, 3
Hale, G. E., 28–29, 30, 32
Hall, J. S., 119
Halley, Edmund, 6, 20, 79, 109
halo, galactic, 115
Harding, G. A., 109
Haro, G., 98, 99
Hartmann, G., 114
Harvard sequence of stellar spectra, 43
Hase, Miss V. F., 92
heliocentric system, 5, 6
Herbig, G. H., 98, 99
Herbig-Haro objects, 99
Hercules cluster, 133, 134
Herschel, Caroline, 129
Herschel, John F. W., 86, 92, 107, 109, 122, 123, 127
Herschel, William, 6, 7, 8, 62, 65, 83, 86, 87, 90, 92, 98, 109, 112, 122, 127, 129
Hertzsprung, E., 57, 123
Hertzsprung-Russell diagram, 56–57, 58, 59, 60
Hesiod, 14
Hiltner, W. A., 118
Hind, J. R., 98
Hipparchus, 15, 16, 17
Homer, 14
hot subdwarfs, 75, 88, 110, 115
Hoyle, Fred, 59, 108, 119, 149, 154
Hubble, Edwin P., 98, 99, 127, 132, 133, 134, 135, 143, 146, 147, 148
Hubble constant, 143, 144, 146, 147
Huffer, C. M., 68
Huggins, W., 86, 87, 127
Hulst, H. C. van de, 116
Humason, M., 129, 143, 146
Huyghens, Christian, 95, 97
Hyades, 14, 102–104, 105, 106, 107, 108, 109
Hydra, 146
hydrogen, interstellar, 99, 100, 104, 116–118, 123, 126
hydroxyl radical, clouds of the, 120
Hydrus, 18
Hynek, J. A., 76

IC 405, 98
IC 434 (Horsehead Nebula), 97
IC 1613, 130
Index Catalogues, 87
Indus, 18
Ingham, M. F., 36
inverse-square law of light, 9
ionosphere, 35

Jeans, J. J., 75, 77
Joy, A. H., 67, 70, 71, 98

Kant, Immanuel, 112
Kappa Crucis, 107
Kapteyn, J. C., 8, 22, 113
Kepler, Johannes, 5
Kepler's star, 74
Kerr, F. J., 116
Keyhole Nebula, 94
Koehler, J. A., 153
Kohlschütter, A., 51
Kowal, C. T., 154
Kraft, R. P., 70, 71, 74
Krüger 60, 48
Krüger 60 B, 47, 48
Krzeminski, W., 71

Lacaille 9352, 22
Lacerta, 97
Lagoon Nebula (M 8), 94, 95
Lalande 21185, 22, 46, 50
Lambda Scorpii, 64
latitude:
  celestial, 15
  galactic, 111
Leavitt, Henrietta, 79, 122
Lemaître, Georges, 147, 153
light curves, 65, 66, 67, 68
light-year, 8
Lindblad, B., 8, 10, 113
Little Dipper (Ursa Minor), 1, 14, 130, 132
Local Group, 126, 131, 132, 143
Local Supercluster (Supergalaxy), 133
Lockyer, J. Norman, 58
Longair, M. S., 154
longitude:
  celestial, 15
  galactic, 111
loop nebulae, 90–92
Lovell, Sir Bernard, 47
Low, F. J., 61
LP 357-186, 49
LP 658-2, 46
luminosities, stellar, 41–42, 44, 47, 52
luminosity function, 132
Lupus, 102
Luyten, W. J., 42, 46, 47, 49
Luyten 789-6, 50
Lyot, B., 32

M 2, 110
M 3, 110
M 13, 79, 109, 110
M 15, 88, 110
M 16, 95
M 17, 94
M 20 (Trifid Nebula), 94
M 27 (NGC 6853; Dumbbell Nebula), 88
M 32, 128, 129, 135
M 33, 129, 130–131, 135, 139, 143
M 42, 97
M 43, 97
M 44 (Beehive; Praesepe), 106, 108
M 51 (Whirlpool Nebula), 115, 116, 127, 135, 138
M 57 (Ring Nebula in Lyra), 87, 88, 89
M 60, 133
M 67, 107, 108
M 74, 135
M 82, 140, 141, 151
M 84, 139

M 87, 71, 133, 135, 138, 139, 141
M 90, 133
M 97 (NGC 3587; Owl Nebula in Ursa Major), 88
M 99, 127
M 100, 133
M 104 (Sombrero Galaxy), 133, 135
magnetic fields, galactic, 118–120
magnetic fields, solar, 29–30, 31, 34
magnetograms, 30, 31
magnetograph, 30
magnitudes, star, 16–17, 41–42
Maia, 104
main sequence, 53–54, 56, 57, 60
  evolutionary significance of, 59–61
Marius, Simon, 127
masses, stellar, 54–56, 59–61, 62
mass-luminosity relation, 54, 56
Matthews, T., 150
Maury, Miss A. C., 64
Mayall, N. U., 129
medium, interstellar, 100, 108
Meltzer, A. S., 66
Mensa, 121
meridian, celestial, 1
Merope, 104
Merrill, P. W., 84, 85
Messier, C., 7, 87, 95, 109, 127
Messier's catalog, 10
Milky Way, 5, 6, 7, 8, 9, 10, 12, 81, 90, 91, 92, 93, 101, 106, 108, 111–120
Milky Way system, 7, 10, 79, 111–120
  disk theory, 111–113
  magnetic field, 118–120
  nucleus of, 120
  rotation of, 113–114
  spiral arms of, 116–118
  spiral structure of, 115–118, 119
  sun's distance from center of, 114
Minkowski, R., 139, 146
Mira (Omicron Ceti), 53, 82, 83
Mizar (Zeta Ursae Majoris), 63, 64
Mizar A, 64
Mizar B, 64
Monoceros, 98, 111
monochromators, 32
Montanari, G., 65
Morgan, W. W., 98, 115
Morris, D., 120
motion, laws of, 2
Mu Cephei, 61, 83
Mu Columbae, 98
Mu Geminorum, 83
Munch, G., 98
Musca Australis, 18

nebulae, 6–7, 8, 10, 86–101, 132
  bright and dark, prominent associations of, 94–97
  Clouds of Magellan and, 122, 123, 126
  dark, identification of, 92
  giant dark, 92–93
  loop, 90–92
  material, nature of, 86–87
  planetary, 86, 87–92
  radio emissions from, 99
  reflection, 93–94
  variable, 98–99
neutron star, 74
New General Catalogue, 87
NGC 188, 107, 108
NGC 205, 128, 129
NGC 1068, 139
NGC 1073, 135
NGC 1300, 135
NGC 1514, 87
NGC 1554, 98
NGC 1555, 98
NGC 1910, 124
NGC 1935 ("Constellation I"), 125
NGC 2261, 98, 99
NGC 2362, 108
NGC 2392 (Eskimo Nebula), 88
NGC 2516, 106
NGC 2859, 135
NGC 3115, 135
NGC 4189, 154
NGC 4651, 139
NGC 4755 (Jewel Box), 106, 107
NGC 5128, 139, 140
NGC 5195, 116, 135, 138
NGC 6522, 114, 115
NGC 6543, 87
NGC 6729, 98
NGC 6822, 131
NGC 6960, 90, 91
NGC 6992, 90
NGC 6995, 90
NGC 7000 (North America Nebula), 92
NGC 7293 (Ring Nebula in Aquarius), 87
NGC 7332, 135
NGC 7635, 89
Newton, Isaac, 2, 62
north celestial pole, 1–2, 14, 16, 21
North Pole, 1
Nova Aquilae, 71, 72, 73
Nova Aurigae, 73
Nova CP Puppis, 71
Nova DQ Herculis, 71, 72, 73
Nova Herculis, 72
Nova Persei, 71, 72, 73
Nova T Coronae Borealis, 71
novae, 61, 132
  bright, 71
  Clouds of Magellan and, 126
  dwarf, 70, 75
  M 31 and, 127–128
  recurrent, 70–71
nutation, 21

Of stars, 75
Omega Centauri (NGC 5138), 9, 109
Oort, J. H., 10, 113, 114, 117, 118
Ophiuchus (Serpent-Bearer), 47, 92, 98
Orion, 5, 13, 14, 52, 53, 93, 98, 111
Orion Arm, 115, 118
Orion Spur, 117
Orion's Belt, 97
Owl Nebula in Ursa Major, *see* M 97

P Cygni, 75
parallax, spectroscopic, 51
parallax, stellar, 7–8, 21, 22, 41, 51

Parengo, P. P., 96
parsec, 8, 41
Pavo, 18
Pearce, J. A., 114
penumbra of sunspots, 24
Perseus, 93, 97, 102, 106, 111
Perseus Arm, 115, 118
Perseus II, 98
phenomena, solar, 32–36
Phoenix, 18
photosphere, 23, 24, 26, 30, 31, 32, 34
Pickering, Edward C., 43
Pigafetta, Antonio, 121
Pigott, Edward, 77
Pisces, 13, 14, 154
Pisces Volans, 18
plages, 32, 34
plane, galactic, 111
Plaskett, J. S., 114
Plaskett's star, 55
Pleiades, 5, 14, 15, 17, 102, 104–106, 108, 109, 115, 124
Pleione, 104–105
Plummer, H. C., 77
Pluto, 13
Polaris (North Pole Star), 1, 16, 17, 77–78, 79
poles, celestial, 1–2, 16
Pollux, 17, 51, 53
poloidal field, 30
Population I objects, 115, 130
Population II objects, 115
populations, stellar, 115, 128
pores, 24
Prentice, J. P. M., 72
"primary minimum," 65
Proctor, R., 102
Procyon, 17, 44, 49, 50, 51, 54
Procyon B, 50
prominences of sun, 28, 29, 33, 34, 35
proper motions, 41, 102
proton-proton reaction, 39
protostars, 110
Proxima Centauri, 11, 47, 50, 51
Ptolemy, 6, 17, 111

quasars (quasi-stellar radio sources; QSS's), 150–154
quasi-stellar galaxies (QSG's), 153, 154

R Andromedae, 57
R Aquarii, 84
R Coronae Australis, 98
R Geminorum, 57
R Leonis, 83
R Monocerotis, 98
Radcliffe Observatory, 9, 123
radial velocity, 8, 41, 158
radiation, synchrotron, 38, 89, 99, 120
radio waves, 35, 37
Ras Algethi (Alpha Herculis), 53, 61, 64, 65, 72
Rayet, G., 73
red-shift, 143, 145, 146, 148
reflection effect, binary stars, 68
refraction, atmospheric, 19–20
Regulus (Little King), 17, 54
relativity, theory of, 39, 147, 148
revolution of the earth, 3, 5–6
Rhijn, P. J. van, 113
Rho Ophiuchi, 92–93
Rho Persei, 83
Rigel, 17, 51, 52, 53, 64
Rigel A, 65
Rigel B, 65
right ascension, 18–19, 20, 21
Roberts, I., 127
Rosette Nebula (NGC 2237), 94, 95
Ross 128, 50
Ross 154, 47, 50
Ross 248, 50
Ross 614 B, 55
Ross 882, 47
Rosse, Lord, 86, 88, 127
Royal Astronomical Society, 11, 35, 36, 43, 44, 52, 57
Royal Greenwich Observatory, 20–21, 26
Royal Observatory (Edinburgh), 15
RR Lyrae, 72, 80
RS Ophiuchi, 70
Russell, H. N., 57, 59
RV Tauri, 81
RW Tauri, 67
Ryle, M., 150
RZ Cephei, 77
RZ Scuti, 67

S Doradus, 122, 124
Sagittarius (Archer), 13, 111, 112
Sagittarius Arm, 115, 118
Sandage, A. R., 59, 108, 140, 141–142, 147, 150, 153, 154
S Andromedae, 74, 127–128
Sanford, R. F., 70
Scheiner, J., 127
Schmidt, M., 150, 151, 153
Schönfeld, E., 22, 102
Schwarzschild, M., 54, 59, 83
Scorpius (Scorpion), 13, 15, 92, 97, 102, 111
Sculptor, 111, 130
"secondary minimum," 65
seconds of arc per year, 41
semiregular variable stars, 81–84, 85
Serpens (Serpent), 93
70 Ophiuchi, 47
Severny, A. B., 30
Sextans, 130
Shajn, G. A., 92
Shapley, Harlow, 9–10, 77, 79, 113, 114, 123, 128, 130
Shklovsky, I. S., 74
sidereal periods, 26
Sigma Octantis, 1
Sirius (Sothis), 6, 14, 17, 42, 44, 47–48, 50, 51, 53, 54, 102, 112
Sirius B, 48–49, 50, 146
61 Cygni, 7, 8, 46, 50
61 Cygni B, 50
Slipher, V. M., 104, 127, 143
solar system, 5, 10, 13, 36, 112
S 147, 91–92
south celestial pole, 1, 121
Southern Cross (Crux), 108, 121
space, geometry of, 148–149
spectra, stellar, Harvard sequence of, 43

spectrographs, 143, 146
spectroheliograph, 28, 32
spectroscope, 8, 63, 86, 127
sphere, celestial, 2, 3, 5, 6, 16, 18, 112
Spica, 15, 17, 51, 54, 64
spicules, 32, 39
SS Cygni, 70, 71
star clouds, 112
stardrift, 102, 113
stars:
  binary, *see* binaries
  brightest, 51
  carbon, 57
  circumpolar, 1, 2, 14
  classification of, 51–58
  exploding, 70–76
  field, 102
  flare, 47, 48, 50
  giant, 53, 54, 56, 57, 59, 60, 61, 81, 83, 109, 115, 118, 124, 125, 129
  luminosities of, 41–42
  magnetic, 30
  magnitudes of, 16–17, 41–42
  main sequence, 53–56
  mapping the, 12–22
  masses of, 54–56, 59–61, 62, 104
  moving systems of, 102
  multiple, 61
  names of, 14, 17
  nearest, 41–50
  noncircumpolar, 1, 3
  pulsating, 77–85
  radio, 47
  rapidly rotating, 75–76
  red dwarf, 45, 46, 47, 51, 53, 55
  RR Lyrae, 80, 110, 114, 115, 124
  sizes of, 5, 45
  spectra of, 43–45
  subdwarf, 57–58
  supergiant, *see* supergiants
  symbiotic, 84–85
  T Tauri, 98
  temperature of, 42–45
  variable, 22, 61, 67, 69, 70, 80, 81–84, 126
  white dwarf, 49–50, 51, 56, 57, 60–61
  Wolf-Rayet, 73–74, 126
star streaming, 113
"steady state" theory, 149–150
Strand, K. A., 46, 96
Stratoscope I, 23, 24
Stratoscope II, 83
Struve, O., 59, 67, 68, 80–81
subdwarfs, 57–58
sun, 3, 6, 23–40, 114
sunspots, 23–30
supergiants, 53, 56, 57, 60, 61, 83, 102, 115, 118, 124, 125, 129, 132
  Clouds of Magellan and, 124, 125
supernova, 60, 61, 74–75, 89, 90, 128, 132
SV Vulpeculae, 79
Sword of Orion, 95, 97
SX phoenicis, 80
synodic period, 26

T associations, 98
T Coronae Borealis, 70–71
T Pyxidis, 70
T Tauri, 98
Taurus (Bull), 13, 93, 98, 102, 104
Taygeta, 104
telescopes, 5, 6, 7, 9, 17, 23, 24, 32, 62, 88
  radio, 47, 100, 126
Tempel, W., 104
thermonuclear reactions, 39, 40, 54, 55, 59, 60
Theta Ophiuchi, 92–93
Theta Orionis, 96
Theta Tauri, 62
Theta-2 Tauri, 103
30-Doradus (Loop Nebula; Tarantula Nebula), 122, 124–125
3C 9, 148, 152
3C 48, 148, 150, 151, 152
3C 147, 152
3C 196, 150, 152
3C 231, 140
3C 273, 139, 148, 150, 151, 152
3C 295, 145, 146, 148, 149, 153, 154
3-kiloparsec arm, 116, 118
Thuban Draconis, 16
Toucana, 18, 78, 109, 121
transverse velocities, 41
Trapezium (Theta-One Orionis), 95–96
Triangulum, 130
Triangulum Australis, 18
Trumpler, R. J., 108
Tucana, 18

U Geminorum, 70
ultraviolet rays, 35
umbra of sunspots, 26
units, astronomical, 8
universe:
  Aristotelian, 4
  expanding, 143–154
  infinite Copernican, 4, 5
  theories of the, 2, 4
universes, model, 146
Uranometria, 17
Ursa Major cluster, 102
UV Ceti, 47, 50
UV Ceti B, 50

Vaucouleurs, G. de, 124, 133
Vega, 16, 51, 54
Veil Nebula (Network Nebula), 90–91, 92
Vela Spur, 117
Virgo (Virgin), 13, 127, 133
Virgo A, 139
Virgo cluster, 133, 139, 153, 154

W Ursae Majoris, 68, 69
W Virginis, 81
Waldmeier, M., 27
Walker, M. F., 73
Weizsäcker, C. F. von, 59
Wickramasignhe, N. C., 119
Wilson, A., 26
Wilson, O. C., 52
Wilson effect, 26
Wolf, C., 73

Wolf, Max, 92
Wolf 359, 42, 44, 50
Wright, T., 111–112
WX Ursae Majoris, 47
WZ Sagittae, 71

X rays, 35, 37, 39

Z Andromedae, 84
Z Camelopardalis, 70
Zeeman, Pieter, 29
zenith, 1
Zeta Geminorum, 77
Zeta Persei, 93
Zeta Scorpii, 54–55
zodiac, 13, 14, 15
zodiacal light, 36
Zwicky, F., 154